Bárbara Nazaré Rocha
Emerson Galvani

Cave microclimate

Bárbara Nazaré Rocha
Emerson Galvani

Cave microclimate

Microclimatic study of the caves of Intervales State Park, SP

ScienciaScripts

Imprint
Any brand names and product names mentioned in this book are subject to trademark, brand or patent protection and are trademarks or registered trademarks of their respective holders. The use of brand names, product names, common names, trade names, product descriptions etc. even without a particular marking in this work is in no way to be construed to mean that such names may be regarded as unrestricted in respect of trademark and brand protection legislation and could thus be used by anyone.

Cover image: www.ingimage.com

This book is a translation from the original published under ISBN 978-613-9-62942-8.

Publisher:
Sciencia Scripts
is a trademark of
Dodo Books Indian Ocean Ltd. and OmniScriptum S.R.L publishing group

120 High Road, East Finchley, London, N2 9ED, United Kingdom
Str. Armeneasca 28/1, office 1, Chisinau MD-2012, Republic of Moldova, Europe
Printed at: see last page
ISBN: 978-620-7-75791-6

Summary

Thanks

To CNPQ - National Council for Scientific and Technological Development - for helping to purchase equipment (process number 485935/2007-4 of the Universal Notice MCT/CNPq 15/2007);

To CAPES - Coordination for the Improvement of Higher Education Personnel - for the grant;

To the Forest Institute/SMA for authorizing the project (SMA process number 260108 - 001.150/ 2008);

To my advisor and friend, Prof. Dr. Emerson Galvani, who was always present in this research;

To Rogério, the lab technician who installed the station and was always on hand when I needed him;

To Mr. Eliseu, a tour guide at Intervales State Park, who looked after the station, took notes on visitors to the park and accompanied me on all the field work;

To the many friends who accompanied me in the field: Sil, Caio, Gui, Pan, Araça, Francês, Pati and others;

To all my family and friends: my base;

To my sister, Mariana, a great companion in the field, with whom I've experienced great adventures;

To my parents, who have always supported me in everything.

Thank you so much!

Summary

The environmental protection of geological resources is an issue of growing interest. Cave exploration is one of the most widespread forms of tourism associated with geological resources. For this reason, its environmental characteristics, such as climate, must be known in order to define management and environmental conservation strategies associated with its tourist use. Thus, the general aim of the research is to detect the main microclimatic changes that occur in the cave environment as a result of tourist visits inside the caves of PEI, SP. The specific objectives were: to characterize the microclimate of the cave environment under natural conditions and to list the physical characteristics of the cave that influence its microclimate.

The methodology consisted of collecting temperature and relative humidity data from nine cavities with autonomous recorders. CO_2 values were also recorded. The variation in climate attributes was analyzed in natural conditions and in the presence of visitors.

The results show that the microclimate of caves tends towards stable air temperatures, high relative humidity and high carbon dioxide concentrations. Caves with flowing rivers and skylights have no impact on their atmosphere as a result of tourism, as gas exchange with the external environment is facilitated. In dry and aphotic caves, with narrow entrances and corridors, the microclimate tends to be stable and is more subject to impacts, especially rises in air temperature. The use of carburetors also generates significant increases in carbon dioxide concentrations.

Keywords: microclimate, cave, environmental impact, air temperature, relative humidity and carbon dioxide.

1. Presentation

My interest in working on the subject of "cave climate" for my master's thesis, entitled "Microclimatic study of the cave environment, Intervales State Park, SP", arose during the Biogeography course I took as an undergraduate geography student at the University of São Paulo. Under the supervision of Professor Dr. Sueli Ângelo Furlan, I carried out fieldwork in Intervales State Park (PEI), where my group collected temperature and relative humidity data from one of the park's caves: the Colorida cave. The unpublished results were presented at the Brazilian Congress of Speleology and pointed to the need for further studies into the climate of this peculiar environment. This work proved that there are changes in the attributes of the microclimate of the Colorida cave, such as an increase in air temperature, in the presence of visitors.

After this work, the invitation to coordinate the climatology team for the Speleological Management Plans for the State of São Paulo consolidated my interest in the subject. The research carried out for the PMEs enriched the results obtained in this dissertation.

"The cave environment *stricto sensu* is considered to be one of the most peculiar and stable in the biosphere. The rocky layer that covers and protects the caves from the sudden climatic variations that occur on the surface gives this environment a series of characteristics of its own, which condition the diversity of animal and plant life that develops within it" (LINO, 2001, p. 203). In this way, caves form a unique environment, with difficult nutrient inputs and a total absence of light in the deeper areas, presenting a specific microclimate that has been studied very little.

The study of this microclimate is important for understanding the factors limiting the development and spatial distribution of life in the cave environment, the development of speleothems and the changes caused by tourism. It can also help to understand the processes of cave formation and global climate change, through a study combined with mineralogical analysis of speleothems.

Visits to conservation units, especially state parks, have become a highly widespread activity in recent decades, leading to an increase in visits to protected natural environments. A large part of this phenomenon can be attributed to the spread of ecotourism, which can have a negative impact on the environment (SANO, 2007).

Cave exploration is one of the most widespread forms of tourism associated with geological resources. For this reason, its environmental characteristics must be known in order to define management and environmental conservation strategies associated with its tourist use.

Due to their physical characteristics, caves are unique environments, with difficult nutrient inputs and a total absence of light in the deepest areas, presenting a specific microclimate.

Lighting and the presence of visitors, among other variables, modify the environmental conditions of cavities, which leads to progressive degradation and even favors the destruction of speleothems. These environmental variations correspond to changes in air temperature and relative humidity, carbon dioxide levels and algae proliferation.

In view of the importance of the subject, the general objective of this research is to detect the main microclimatic changes that occur in the cave environment as a result of tourist visits in order to subsidize management actions inside the caves of the Intervales State Park, SP. The specific objectives were: 1) to characterize the microclimate of the cave environment under natural conditions and 2) to list the physical characteristics of the cave that influence its microclimate.

To do this, the study must first record the cave's natural environmental conditions, which will allow us to detect the environmental disturbance caused by human groups inside the cave. Thus, the spatial and temporal variations of the attributes should be analyzed according to the number of visitors and length of stay in the caves, the natural ventilation capacity of the cave, the degree of dripping and the size and shape of the rooms.

In order to achieve the proposed objectives, some hypotheses guided the development of the work.

- The microclimate of caves, in a natural situation, must show delays in the assimilation of external atmospheric conditions, and its attributes do not vary in the aphotic zones;
- Tourist visits, through the use of artificial lighting and human body heat, can alter the air temperature in the cave environment, which consequently changes the relative humidity;
- Cave tourism alters the concentration of carbon dioxide in the atmosphere of the cave environment as a result of lighting using carbide.

This work is structured in the following chapters: *1- Presentation*, where the objectives, hypotheses and justification *of* the work are outlined; *2- Literature Review*, covering citations and summaries of the main works consulted; *3- Theoretical-conceptual framework*, with the guiding theories of the work and research procedures; *4- Study area*, characterizing the Intervales State Park and the caves researched; *5- Results and discussion*, with the results and analysis of the data; *6- Conclusions and final considerations*, with the conclusions and notes on the work, *7- Bibliography*, where the works consulted are presented and 8- *Appendices*.

It should be noted that this work was made possible thanks to the financial support granted by CNPQ - National Council for Scientific and Technological Development - for the purchase of equipment (process number 485935/2007-4 of the MCT/CNPq 15/2007 Universal Call for Proposals), and by CAPES - Coordination for the Improvement of Higher Education Personnel -, with the granting of a scholarship. The research in the park area was authorized by the Forestry Institute (SMA process number 260108 - 001.150/2008).

2. Literature review

A natural underground cavity is defined as "any subterranean space penetrable by humans, with or without an identified opening, (...) including its environment, its mineral and water content, the biotic communities found there and the rock body in which they are inserted, provided that it was formed by natural processes, regardless of its size or the type of rock it contains" (BRASIL, 2004). In short, we will understand in this work that "caves are natural cavities with dimensions that allow human access" (Karmann, 2003).

The microclimate of the cave environment is still little known. Most studies on this subject involve an analysis of just two attributes: temperature and relative humidity.

By microclimate we mean the layer of air close to the surface.

"The layer of air below the agreed height of about 2m is called the layer of air near the ground. (...) The closer we get to the ground surface, the greater the friction and therefore the lower the wind speed and, consequently, also the mixing of the air that would make it possible to compensate for the difference in small spaces" (GEIGER, 1990).

Carvalho (2004) carried out a study of the underground microclimate in the Olhos D'àgua Cave in Castro, PR. In this work, temperature and relative humidity measurements were taken using psychrometric sets, which makes it impossible to detect any influence of human presence in the cave environment, as the measurements can only be taken in the presence of the researcher.

A microclimatic study of the Ubajara Cave, CE, showed that the temperature and relative humidity of the air at the entrance to the cave varied more than those inside, with greater fluctuations in the dry season compared to the rainy season (VERiSSIMO *et al*, 2003). This study also showed that the lighting *spots* contribute to raising the air temperature in the smaller rooms of the cave. For this reason, many caves have already replaced the *spots* with cool lamps.

Longhitano *et al* (2006) and Rocha *et al* (2006) carried out a preliminary characterization of the microclimate of the cave environment through a case study in the Gruta Colorida in Intervales State Park, SP. These studies used five temperature and relative humidity loggers placed along the upper gallery of the cave (from the entrance to the end of the corridor) over a two-day period. There was a delay in assimilating changes in the weather outside as one entered the cave. A variation in the air temperature inside the cave was also detected as a result of human presence.

The increase in air temperature due to tourist visits was also pointed out in the studies by Viana Jùnior (2002) and Sànchez-Martos *et al* (2002). In the former, changes of more than 2°C were recorded in the Santana cave of the Alto Ribeira Tourist State Park (PETAR), SP, in the presence of three visitors, without the use of a carburetor. The second study was carried out in the Cueva del Agua de Iznalloz cave in Granda, Spain. Sànchez-Martos *et al* (op. cit.) concluded that the magnitude of changes in air temperature is directly related to the number of visitors and their time in the cave and that the cave recovers thermally within a few hours of the visit.

Scaleante (2003) concluded that the use of carburetors is the main factor responsible for the rise in air temperature in cave environments. In his dissertation, he observed that the passage of 310 visitors through the Salao do Encontro, in Santana Cave, SP, did not cause an increase in air temperature, while seven visitors with carburetors caused an increase of 4.5°C. Furthermore, when the environment is impacted by the presence of visitors using carbide lighting, the cave takes longer to recover thermally.

According to studies carried out in the painting room of the Altamira cave in Spain, the heat dissipated by the human body varies from 82W to 116W (VILLAR et al, 1984 *apud* SCALEANTE, 2003).

Sano (2007) points out that "caves are sensitive environments to the presence of visitors inside them, because there is a change in temperature, with the heat emitted by the human body, humidity and air composition due to human breathing and the gas emitted by the carburetors, which stick to the speleothems and take away the natural beauty with their black residue. In addition, the excessive number of people walking inside the caves has an impact on the soil, which becomes compacted, and in some cases they can leave garbage, such as leftover food and bottles."

Also, "the mere presence of humans in cave environments, provided that it is greater than what the system is capable of absorbing, causes irreversible impacts on the biota, the rock mass and the internal formations" (SCALEANTE, 2003).

According to CONAMA Resolution 001 (BRAZIL, 1986), environmental impact is defined as "any alteration of the physical, chemical and biological properties of the environment, caused by any form of matter or energy resulting from human activities which, directly or indirectly, affects: the health, safety and well-being of the population; social and economic activities; biota; the aesthetic and sanitary conditions of the environment; the quality of environmental resources". Along the same lines, Moreira defines environmental impact as "any alteration to the environment in one or more of its components - caused by human action" (1992, p 113). Or even "the change in an environmental parameter, over a given period and in a given area, that results from a given activity, compared to the situation that would have occurred if that activity had not been initiated" (WATHERN, 1988, p.7 *apud* SANCHEZ, 2008). Impact also refers to the capacity of the environment to absorb it.

Thus, impacts can be positive or negative, reversible or permanent. Degradation of environmental quality, on the other hand, refers only to negative changes and is defined as "the adverse alteration of the characteristics of the environment" (BRASIL, 1981).

Environmental impacts result from an environmental aspect, which is the "mechanism through which a human action causes an environmental impact. The same action can lead to several environmental aspects and therefore cause several environmental impacts. In the same way, a given environmental impact can have several causes" (MOREIRA, 1992, p. 33 *apud* SANCHEZ, 2008).

Human activities -> environmental aspects -> environmental impacts

In the cave:

Tourist visits -> increase in available energy -> rise in air temperature

Environmental impacts can be irrelevant or significant. "Significant is anything that has meaning; it is synonymous with expressive. But the term significant environmental impact should be understood as meaning considerable, large enough, or even important. This definition, however, does not solve the problem because significant impact is a subjective term. And it could hardly be otherwise, since the importance attributed by people to environmental changes called impacts depends on their understanding, their values, their perception" (MOREIRA, 1992, p.111 *apud* SANCHEZ, 2008). For this reason, it is common to make mathematical correlations and assign weights to the variables analyzed. In this work, the impact that exceeds the amplitudes found in a natural situation is considered significant.

Tourism in caves does not only have an impact on the temperature regime. The cave microclimate is quite stable (LINO, 2001). Therefore, small climatic changes can cause major changes in the cave environment. Tourism, by widening corridors, alters circulation; lighting raises the air temperature and allows the growth of algae and a consequent increase in the rate of carbon dioxide; human respiration also helps to create a greenhouse environment. The increase in CO_2 can affect the chemical balance of speleothems. A 1°C increase in temperature results in an eightfold increase in the capacity to concentrate water vapor, decreasing the relative humidity of the air and increasing the concentration of carbon dioxide at that instant for that temperature (GILLIESON, 1996).

The increase in the concentration of carbon dioxide in the cave environment as a result of human presence was detected in the study by Linan *et al* (2006). The rate of CO_2 inside a cave is conditioned by the cave's natural ventilation capacity. Ventilation is a function of the temperature difference between the external and internal environments. Therefore, if there is an increase in the internal temperature of the cave as a result of human presence, there will be variations in circulation and, consequently, in carbon dioxide rates. In addition, CO_2 is exhaled by humans in the respiratory process.

Palomba (1995), after studying the Pozzi della Piana cave in Italy, concluded that several factors contribute to the increase in the concentration of carbon dioxide inside the cave. The factors can be related to the physiology of the cave, such as the production of oxidizing organisms from the organic material present in the substrate and the respiration of the fauna, or originate from tourist visits, such as the combustion of acetylene produced by the carburetors and the respiration of visitors.

3. Theoretical-conceptual framework

The geosystemic theory was adopted as the theoretical basis for this research. This is widespread in geographical studies and is appropriate for this study because, according to Sotchawa (1977):

- It is pertinent and suitable for systematic organization and the search for synthesis;
- It is concerned with the relationship between dynamic variables and genesis;
- It allows us to deal with the principles of interdisciplinarity, synthesis, multi-scalar and dynamic approaches;
- It promotes the recognition of spatial units with elementary, relational and dynamic characteristics that are similar to each other, even including the anthropic;
- Enables the breakdown of operational planning units;
- Systems can be defined as sets of elements with different variables and characteristics, which maintain relationships with each other and the environment.

This theory considers nature as an open and hierarchically organized dynamic system that can be delimited or spatially circumscribed in its three-dimensionality (RODRIGUES, 2001: 73). Geosystems represent the spatial organization resulting from the interaction of nature's physical components (systems), including climate, topography, rocks, water, vegetation and soils, among others, all of which may or may not be present.

Geosystems are twofold, allowing, on the one hand, analysis of the homogeneous structure, which characterizes the geomer, and, on the other, the integrative qualities of geosystems, which characterizes the geocore. A geomer is understood to be the homogeneous structure of geosystems, while geocores are heterogeneous structures, but allow for an integrated analysis to better understand the dynamics of systems (SOTCHAWA, 1977).

Dynamics classifies geosystems according to their successive states, and can assume or propose hypotheses about their future dynamics, a fundamental characteristic for application and planning. Thus, the predictive nature of this theory is one of its main applications.

Based on this theory, data can be systematized, starting with modeling and measuring the exchanges, circuits and balances of matter and energy in systems and subsystems.

Furthermore, understanding the object of study as a geosystem allows for an environmental assessment of it, since caves are systems of dual structures, with a flow of matter and energy, and can be spatialized.

The methodological basis of this research is analytical reasoning, which relates facts (parts) to explain the context (whole). Among the different types of analysis, the hypothetical-deductive approach was chosen, with hypotheses being formulated as provisional answers, these being the analytical variables of the work. The reasoning was also organized on an empirical basis, by means of controlled experimental analyses.

The technical procedure for the study took the form of systematic data collection through fieldwork.

The caves surveyed were: Fendao, Maozinha, Minotauro, Jane Mansfield and Santa, from the Bocaina/Lageado grouping (PEI) and Fogo, Meninos, Detrâs and Tatu, from the Sede nucleus. The choice of caves followed the proposal of the Speleological Management Plans, which consider these caves to be the most visited in the park. They are therefore the subject of a public use plan and a prohibition requirement.

The field work allowed for the installation and removal of temperature and relative humidity sensors and the collection of carbon dioxide concentration data. At this stage, three thermohygrometers were installed in each cave, one at the entrance (characterizing the transitional microclimate), another in a room within the traditional visitation circuit and the third in an unvisited section, but peripheral to a visitable area. The equipment used to monitor the cave was sensitive and accurate enough to record small variations. It should be noted that the use of these instruments was only possible with the support of the National Council for Scientific and Technological Development (CNPQ).

The table below (table 3.1) shows the equipment used in each of the clusters studied and its characteristics.

Table 3.1: Equipment used

Instrument	Brand	Model	Precision	Accuracy
thermohygrometers	Onset	Stow Away	0,1	0.2°C (T) and 2.5% (RH)
Carbon dioxide sensor	Tracom	TEL-7001	0,1	50 ppm

Organization: Bàrbara N. Rocha, 2010.

The monitoring of each cave was carried out over a period of one week (seven days) with a time resolution of the thermohygrometers of six minutes. The aim was to show the natural variations in climate attributes (periods with no visitation - control situation), and the impact of visitation. For the second case, simulated visits were made with groups that reproduced the number, length of stay and behavior of traditional visitors. Field work to install and remove the devices took place between February and April 2009.

Visits to the cave were carried out using a questionnaire (appendix 8.1), which included: the date and time when visitors entered and left, the number of visitors and the number of flashlights and lanterns used by tourists.

Carbon dioxide data was collected at various points in each cave (a minimum of ten samples), covering the entire traditional tour circuit.

After collecting the data, thermal, hygrometric and carbon dioxide (CO_2) profiles were created for each cavity. Working with profiles makes it easier to visualize the changes in the parameters analyzed, making it possible to establish correlations between changes in each parameter and the presence of visitors.

In addition to the profiles, the temperature and relative humidity were also analyzed using statistical parameters (mean, maximum, minimum, amplitude, mode, variance and standard deviation), with emphasis on the amplitudes recorded during the periods with visits. The study of the mode, the value that appears most frequently in the samples, is important in this work, as it shows the temperature and humidity values expected to be found in caves where there is microclimatic stability. The median also helps to detect stable climates, because when the data distribution is symmetrical, the mean and median coincide.

Line graphs were generated from the carbon dioxide data, showing the variation of the parameter from the cave entrance to its end.

An automatic weather station was installed outside, near the Intervales State Park headquarters, consisting of sensors that record air temperature, relative humidity, wind direction and speed, rainfall and solar radiation. The use of this equipment is justified for climate control of the external environment and comparison with the data obtained inside the cave (figure 3.1).

Figure 3.1: Weather station installed at P.E.I.

By Barbara N. Rocha, 2009.

All the data obtained was calibrated for use. Before being installed in the caves, the temperature and humidity loggers were kept together, under the same climatic conditions, for later comparison and calibration. The station, on the other hand, was installed for over a month next to another one from USP's Geography department, and provided very reliable data.

Finally, the parameters collected were analyzed in an attempt to understand the natural dynamics of the cave microclimate and the impacts of visitation on the cave atmosphere, which can support and/or point to guidelines for establishing the carrying capacity of each cave and environmental zoning.

It should be noted, however, that the methodology for this study has some limitations, as the period of climate monitoring does not allow us to understand the natural seasonal changes in the climate of the cave environment.

In order to better understand the results, as well as to characterize the area and the theme, bibliographic sources were used in the fields of geographic climatology, speleology, karst geology, karst geomorphology and biogeography. Cartographic sources included thematic maps of the Intervales State Park and plans of the caves studied.

The techniques adopted in this work followed the following sequence:

1) Bibliographic and cartographic research: the bibliographic survey enabled the

formation of a theoretical-methodological basis and an understanding of the microclimatic dynamics of the cave environment, while the cartographic survey helped to characterize the area studied.

2) Preparing the fieldwork: helped choose the caves studied and characterize them.

3) Field data collection: allowed the installation and removal of temperature and humidity recorders and the recording of atmospheric carbon dioxide data.

4) Field data analysis: analysis and processing of data obtained in the field in order to produce graphs and statistical and correlation analyses.

5) Writing the final paper: presentation of the progress of the work and the results obtained.

4. Study area

The Intervales State Park (PEI) is located in the southern region of the state of São Paulo, 270 km from the capital. It is located in the municipalities of Guapiara, Ribeirao Grande, Sete Barras, Eldorado and Iporanga, on the watershed of the Paranapanema and Ribeira de Iguape basins. It covers the former Banco do Estado de São Paulo - BANESPA farm, acquired by the Forestry Foundation - SMA, with a total area of 41,705 hectares (SAO PAULO, 2009).

Figure 4.1: Location of P. E. Intervales.

Source: SAO PAULO, 2009.

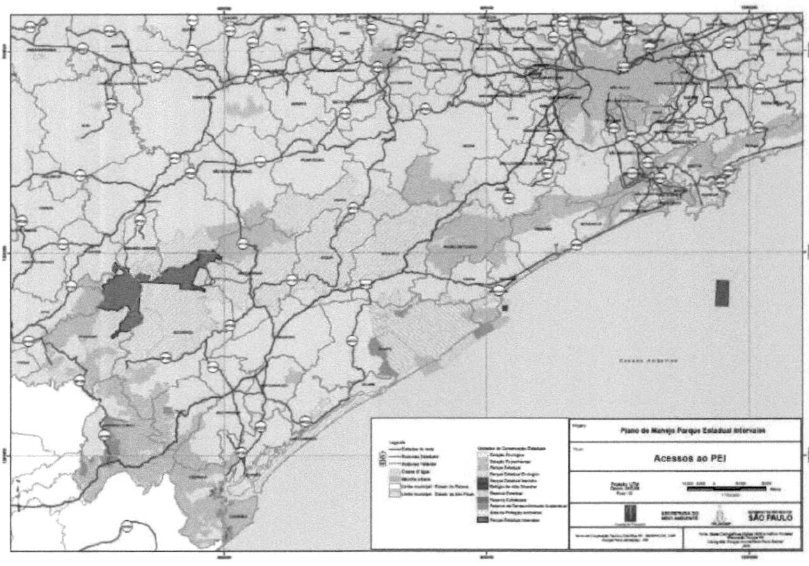

The park's history begins in the second half of the 20th century, when the Companhia do Incremento Rural do Altiplano Paulista settled in the Paranapiacaba mountains to implement an agricultural project that was unsuccessful, losing its land to the former Banco do Estado de São Paulo. In the 1970s, in order to make economic use of the area, BANESPA carried out infrastructure work, such as building roads, a small village with basic sanitation (the current headquarters) and surveillance bases. A palm heart processing factory was set up in the Sede region, where mechanical and carpentry workshops and a warehouse currently operate.

In the 1980s, with the decreeing of the Serra do Mar Environmental Protection Area, the Intervales farm was incorporated into it, passing to the administration of the

Forestry Foundation in 1987. In 1995, the Intervales State Park was legally created (decree no. 40.135, of 08/06/1985).

The term state park can be understood as a "land and/or water area for environmental protection, endowed with environmentally relevant natural attributes to be preserved on a permanent basis. Its basic objective is the *in situ* conservation of biodiversity, its functional dynamics and fundamental ecological processes, as well as its landscape, cultural and historical attributes, intended for scientific, educational, recreational and ecotourism purposes, in accordance with the guidelines of its management plan, safeguarding the area from any alterations that may distort it *in* order to guarantee the preservation of its ecosystems" (Intervales State Park: environmental management plan - phase 1, p. 09).

According to Decree No. 40.135 of June 8, 1995, which created the Intervales State Park, the area is intended to protect water sources, speleological sites and the slopes of the Serra do Mar / Paranapiacaba, covered by the Atlantic Forest. This domain has been awarded the title of Biosphere Reserve by UNESCO and is home to dozens of endangered species, which can only survive in large forested territories (SAO PAULO, 1995).

According to Ross and Moroz (1997), the PEI is geomorphologically part of the Escarpment/Sierra do Mar and Coastal Hills unit, which consists of a strip of steep slopes bordering the Atlantic plateau from the Bocaina plateau region, on the border with the state of Rio de Janeiro, to the Vale do Ribeira de Iguape region. Denudational landforms predominate, basically consisting of escarpments and ridges with sharp tops and convex tops, with valleys ranging from 80 to more than 160 meters and interfluvial dimensions ranging from less than 250 to 3,750 meters. Altitudes range from 20 to 1000 meters and the predominant slopes are over 40%. The lithology of this morphological unit is basically made up of gneisses, migmatites, mica schists and granites.

In this highly dissected relief unit, the drainage has a dendritic pattern, adapted to the directions of the structures that are related to faults, fractures and lithological contacts. As it is a unit with very intense dissection forms, with steep-sided valleys, high drainage density and very steep slopes, this area has a very high level of potential fragility and is subject to aggressive pluvial-fluvial erosion processes and spontaneous and induced mass movements (ROSS, MOROZ, 1997).

The regional geology of Intervales State Park is made up of three major rock units: units related to the crystalline basement, units of metasedimentary rocks (including carbonate rocks) and intrusive granitoids (ROSS, MOROZ, op. cit.). In the Intervales State Park area, metasedimentary rocks of various compositions predominate. The metasediments represent a depositional sequence related to a deep-water marine environment, moving on to shallower marine sediments, from the oceanic platform to the continental shelf.

As it is located in a mountainous area, the relief of P. E. Intervales is strongly dissected. Intervales is strongly dissected, forming a transition between the elevated terrain of the Guapira Plateau and the lowlands of the Ribeira Valley, with altitudes varying between 1095 and 60 meters.

The northern and northeastern parts of the park form an important watershed for two major river basins, the Ribeira de Iguape and the Paranapanema. Most of the park's rivers drain into the Ribeira, except for the northwestern portion which flows into the Paranapanema. In the mountain and escarpment sectors, the rivers have deep valleys and waterfalls.

Most of the rugged front of the Paranapiacaba mountain range is granite, while the prominent ridges are usually quartzite and the higher reliefs are supported by limestone. The phyllites and schists form low reliefs and are home to the main subsequent rivers.

According to the geomorphological compartmentalization adopted by Ross and Moroz (1997), most of the park area falls within the Atlantic Plateau morphocultural unit and a small area within the Lower Ribeira Depression. The Atlantic Plateau is a constituent part of the Atlantic Orogenic Belt with high and very dissected relief, predominantly supported by crystalline rocks of Precambrian age.

The relief of the park can be divided into three main compartments: the Guapira Plateau, the Ribeira Turvo Plateau and the Lower Ribeira Plateau. The first compartment is located in the northern region, with predominant altitudes between 800 and 1000 meters, representing the highest part of the park. The Ribeira Turvo plateau occurs over almost the entire length of Intervales, with altitudes of between 200 and 800 meters. Finally, the southeastern portion is home to the Lower Ribeira Plateau, with altitudes ranging from 60 to 190 meters.

The main types of soil found in the park are: Litholic Neosol, Fuvial Neosol, Cambissolo Hàplico, Red-Yellow Latosol, Bruno-Yellow Latosol and Organosol associated with gleissolo. There is a predominance of Litholic Neosols and Cambissolos Hàplicos associated with granite, phyllite or limestone. This is followed by Red-Yellow

Latosols associated with granite and Bruno-Yellow Latosols associated with phyllite, both of which have a clay texture and are associated with Red-Yellow Argisols. The Fuvic Neosols are associated with phyllite, limestone, marble, diabase and siltstone, mica schist and alluvial material. Organosols and Gleissols associated with mica schist and alluvial material are less common (SAO PAULO, 2009).

The region's predominant climate is subtropical highland (Cwb), according to the Koppen climate classification, with high rainfall in summer and no dry winter, with rainfall throughout the year.

The climate of the Ribeira de Iguape River Valley region is classified, according to Monteiro (1973), as having a humid climate on the eastern and subtropical side of the continents dominated by a tropical maritime mass, controlled by tropical and polar masses. Rainfall follows the layout of the relief and the orientation of the coast in relation to the regional atmospheric circulation currents.

"While the rhythmic unit is characterized by the higher rate of penetration of polar masses and frontal passages, the quantitative distribution of rainfall varies according to the topography. Here we have the coastal stretch of São Paulo exposed in winter to sensitive and more frequent drops in temperature. Even in summer, its polar participation rate is the highest in the state. (...) The wettest part of the coast and especially the exposed face of the isolated massifs gives way to a less humid part along the course of the Ribeira, increasing again on the slopes of Paranapiacaba. Topographical variations allow for a great multiplication of local climates." (MONTEIRO, 1973).

Sant'anna Neto (1990: 137) arrived at results that were in agreement with Monteiro's (1973) characterization of the region's climate.

"The rainfall genesis of the São Paulo coastal zone is associated with frontal activity. (...) The southern sector of the São Paulo coastal zone is clearly under the control of the polar masses (PA and PV). (...) The relief, both due to its altitude, but mainly due to its disposition in relation to the predominant direction of the atmospheric systems coming from the southern quadrant, assumes in the São Paulo coastal zone, a peculiar fact in the rainfall genesis."

According to Monteiro (1973), this region is characterized by a rhythmic unity with a significant influx of polar masses and frontal systems, even in summer. The high slopes of the Serra de Paranapiacaba act as a barrier to the advance of air masses coming from the south of the continent and the Atlantic Ocean, which causes the clouds to be pushed to the top of these cliffs and condense, thus causing a lot of rain and fog.

The total annual number of rainy days in the park varies from 125 to 150 (SAO PAULO, 1994), with average rainfall ranging from 1650 to 1900 mm per year. January, February and March are the wettest months, while June, July and August have the lowest rainfall (SAO PAULO, 2009).

Due to the high rainfall, relative humidity is rarely below 40% and the park average is 83.4%. Land use and aspects of vegetation cover are the main controls on this attribute: areas with a dense, closed canopy have the highest relative humidity levels (SAO PAULO, 2009).

The average annual temperature in the park is 18.4°C, ranging from 14.3°C in the coldest month (July) to 22°C in the hottest month (February). In the hottest months, maximum temperatures are around 34°C; in the coldest months, absolute minimum temperatures can reach 4°C at the highest altitudes (SAO PAULO, 2009).

Altitude as an important climatic control on the São Paulo coast has already been extensively studied by various authors, such as Conti (1975), Sant'anna Neto (1990), Milanesi (2003), Pelegati (2005) and Armani *et al* (2005 and 2006). Sant'anna Neto demonstrates that "of all the geographical location factors, none seems to have a greater influence on the spatial distribution of rainfall in São Paulo than altimetry" (SANT'ANNA NETO, 1995: 53).

The orography intensifies the rainfall on the mountain slopes to windward of the predominant flow from the lower troposphere and reduces the totals to leeward. The Atlantic Ocean also plays a fundamental role in transferring heat and humidity to the surrounding atmosphere.

According to Monteiro (1975), the spatial units of climate are closely linked to geomorphological units, because:

"Climate is positioned in the concrete, three-dimensional space of the earth's surface through what constitutes its framework - the shapes of the terrain (p.131). (...) The Serra do Mar not only separates climatic units, between the hot and humid coastal lowlands and the plateaus, but also influences the conditions of settlement and land occupation." (MONTEIRO, 1976: 23).

Gutjahr (1993) points out that the Ribeira de Iguape region has almost no dry periods or months and that rainfall is largely conditioned by the relief. The author also showed the importance of altitude in the spatialization of air temperature, as well as rainfall, for the region.

"Thermal variations also reflect the influence of geomorphological compartmentalization and the dynamics of atmospheric circulation in their spatial responses represented by the variation in rainfall (...) The most expressive cooling is found on the plateaus and mountains and the greatest warming is linked to the driest areas" (p. 51).

The predominant vegetation cover in the P. E. Intervales is the Dense Ombrophilous Forest (Tropical Rain Forest), the most exuberant of the Atlantic Forest ecosystems, with a great deal of endemism. Although the vegetation of the PEI shows exuberant vegetative vigor, it is mostly Secondary Forest derived from Montane or Submontane Ombrophilous Dense Forest.

According to Decree No. 750/93, the Atlantic Rainforest encompasses a diverse mosaic of forest ecosystems with very different floristic structures and compositions, great biological diversity and high degrees of endemism of flora and fauna, accompanying the diversity of soils, reliefs and climatic conditions of the vast region where it occurs. It ranges from dense ombrophilous forest (figure 4.2) to semi-deciduous forest and mixed forest.

Figure 4.2: Dense ombrophilous forest on the Gruta Colorida trail - PEI.

By George A. Longhitano, 2004

This domain has been considered one of the most remarkable in terms of ecological value, as it is home to typical species and biological attributes that are unique on the planet. This biodiversity corresponds to the variability of living organisms of all origins and the ecological complexes of which they are a part. It also includes diversity within species, between species and of ecosystems, and can be explained by the great altitudinal variety, and by the overlapping of two domains, the tropical Atlantic and the Araucaria Forest, which constitutes the Mixed Ombrophilous Forest.

In the park area, as the forest moves inland towards the north, it receives a greater contribution from the plant species of the seasonal semi-deciduous forest. Factors such as relief, lithology, soils and climate play a major role in the diversity of vegetation physiognomy.

The lithology of the Intervales State Park, with the presence of carbonate rocks, especially limestone, combined with the humid climate, rich drainage and lush vegetation of the region, allowed the development of the caves present in the Park.

According to CONAMA resolution 347, of 10/09/2004, which provides for the protection of speleological heritage, an underground natural cavity is defined as "any underground space penetrable by human beings, with or without an identified opening, popularly known as a cave, grotto, lapa, burrow, abyss, furna and hole, including its environment, its mineral and water content, the biotic communities found there and the body of rock in which they are found, provided that they were formed by natural processes, regardless of their size or the type of rock they are found in" (BRASIL, 2004).

Caves are the result of the action and circulation of water over soluble rock. According to Lino (2001), in order for karst relief to develop or for karst forms such as caves to exist, certain basic conditions must exist:

1) There must be a considerable thickness of rock in the area that is reasonably soluble in slightly acidic water (carbonic acid) from rainfall and surface streams. The rock must be laid down in thick, continuous blocks, be compact and crystalline and have a high degree of diaclasation (joints and fractures) and bedding (rock layers), especially in thin strata. The type of rock best suited to these conditions is limestone;

2) The rock package must have an "available relief", i.e. an elevation of the area above sea level and considerable portions of it above the water table, which allows the free circulation of underground water and the complete development of caves;

3) Climatic conditions associated with vegetation cover also make a decisive

contribution to the development of the karstification process and the formation of caves. In humid regions, the presence of dense vegetation contributes to the acidification of the water and, consequently, to the dissolution of the rock.

4) As described above, all these conditions are present in the EIP.

According to Karmann (2003), carbonate caves are formed from the dissolution of calcite (calcium carbonate - $CaCO_3$) by carbonic acid (H_2CO_3). Carbonic acid is formed when rainwater percolates through the region's soil, which is rich in organic matter and whose decomposition generates carbon dioxide (CO_2).

$$H_20 + CO_2 \rightarrow H_2CO_3$$

water carbone dioxide carbonic acid

Acidified water penetrates through the cracks in the limestone, attacking the rock and producing calcium bicarbonate [$Ca(HCO\ )_{32}$], which is soluble and easily transported by water. The cracks slowly widen, forming underground cavities.

$$H_2CO_3 + CaCO_3 \rightarrow Ca(HCO_3)_2$$

carbonic acid calcium carbonate calcium bicarbonate

When the cave is already formed, the transported calcium bicarbonate emerges on the ceiling of the cave until it reaches sufficient volume and weight to fall. At this point, carbon dioxide gas is released and calcium carbonate precipitates, giving rise to the stalactites (figure 4.3).

$$Ca(HCO_3)_2 \rightarrow CaCO_3 + H_2O + CO_2$$

calcium bicarbonate calcium carbonate water carbon dioxide

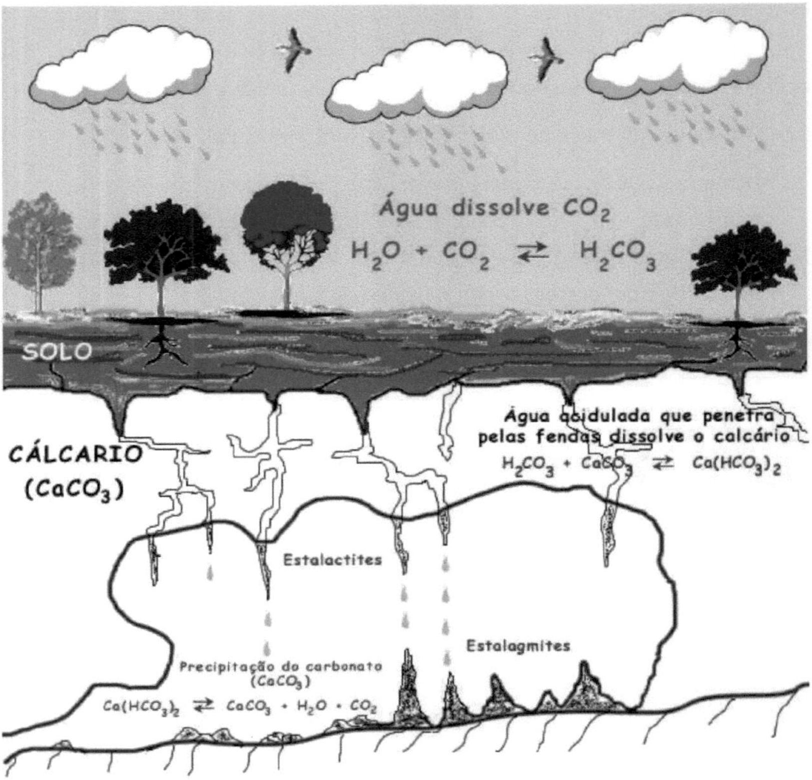

Figure 4.3: Formation of limestone caves.

Source: http://www.cdcc.usp.br/quimica/ciencia/cavernas.html

Thus, geological, geomorphological, climatic and vegetation factors act together in the process of speleogenesis. This is a dynamic process and the caves that originate from it should not be understood as finished products, but rather as subterranean components of a continually evolving relief.

Due to these conditions, the Ribeira Valley is the area of greatest speleological importance in the southeastern and southern regions of Brazil, covering portions of the states of Paranà and Sao Paulo (LINO, 2001).

Intervales State Park has dozens of caves, formed from the dissolution of limestone. The map below (figure 4.4) shows the most visited caves in the park, which are the subject of this study.

Figure 4.4: Location of the caves studied in

Intervales State Park.

By George A. Longhitano, 2010. Source: EKOS Institute, 2009.

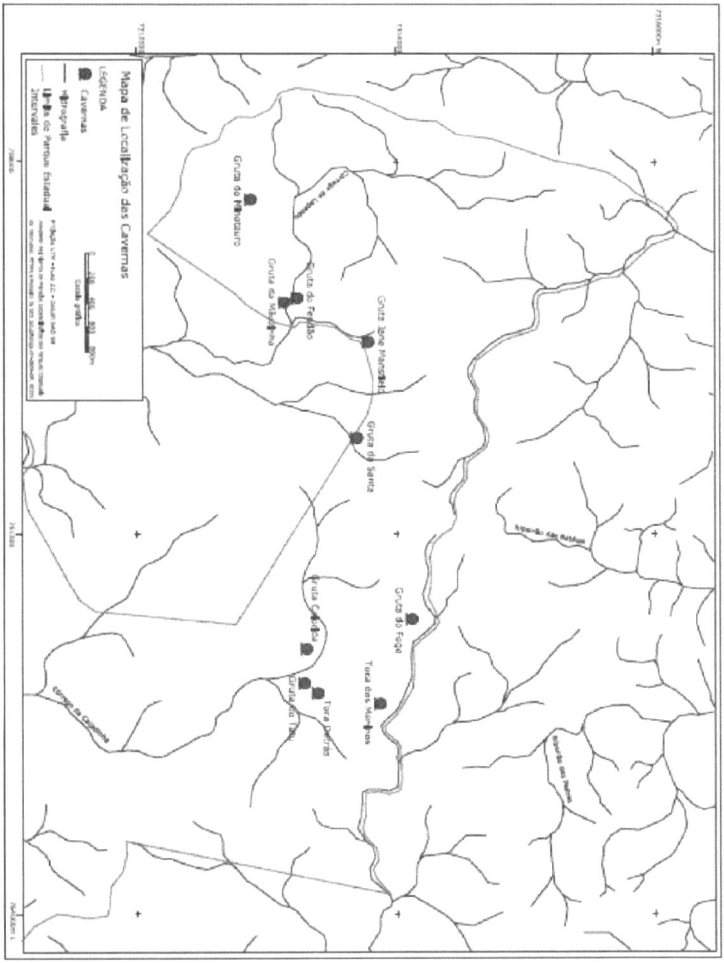

Near the park's headquarters are the Tatu, Detràs (or Cipó), Fogo and Meninos caves. The caves in the Sede nucleus are small (table 4.1) and, consequently, not very popular with visitors.

Table 4.1: Approximate length of the caves studied

Head Office	Bocaina/ Lageado grouping

Grotto	Length (m)	Grotto	Length (m)
armadillo	32	Cleft	1120
Detràs	60	Maozinha	54
Fire	126	Jane Mansfield	500
Boys	30	Santa	49
		Minotaur	560

Organization: Barbara N. Rocha, 2010. Source: SAO PAULO, 2009.

The Tatu cave has few formations and bats and opossums can be seen in it. It has skylights and a running river. The visit circuit is only 25m long.

The Cipó cave is dry, i.e. it is not cut by a river, and has formations such as stalagmites and stalactites. It is inhabited by bats and is approximately 60m long.

The Meninos burrow is also dry and the smallest of those studied - it is less than 30m long. It is accessed via a narrow entrance near the Mono Carvoeiro guesthouse. It is inhabited by bats and opossums.

The Cave of Fire is approximately 130m long and has a small river with calm waters. There are no speleothems inside.

In the Bocaina-Lageado grouping, the Santa, Jane Mansfield, Minotaur, Fendao and Maozinha caves were studied. Of these, the Fendao cave, which is over a kilometer long and has a flowing river full of waterfalls, is the most visited. The Jane Mansfield, Santa and Minotauro caves, which are medium-sized, have average visitation. Only Maozinha, the smallest cave in this group, receives few tourists. The Maozinha cave is next to the Fendao cave; it doesn't have a river, but it does have some speleothems and is covered in mud.

The Jane Mansfield cave has a narrow entrance and is cut off by a flowing river. Here you can find stalagmites, stalactites, curtains, travertines and columns.

The Santa cave has some stalactites and travertine dams on the floor. Its dry environment provides a place for bats and opossums. An image of Our Lady of Lourdes stands at the entrance, next to an altar, giving the cave a religious tourist use (figure 4.5).

Figure 4.5: Entrance to Santa's Cave.

By Bárbara N. Rocha, 2007.

Finally, the Minotaur Cave is named after its narrow, Iabyrinthine corridors. It is approximately 560m long and is crossed by a small river. There are formations of stalagmites, stalactites and curtains.

To make it easier to understand the cave microclimate, each of these caves will be better described in Chapter 4 - Results.

Inside these caves are deposits of sediment accumulated on their floors and recesses. There are also mineral deposits that cover and develop from the ceilings and walls of the cavity. Considering their origins and depositional processes, these sediments are divided into two large groups, according to Lino (2001):

J Clastic sediments: made up of collapsed blocks and detrital material of generally superficial origin transported into the cave by water currents, wind or gravity. There are also frequent silty-clayey or even sandy fractions from the alteration and subsequent erosion of the cave's underlying rock.

J Speleothems: from the Greek *spelaion* (cave) and *thema* (deposit). They are defined as mineral deposits in caves that are basically formed by chemical processes of dissolution and precipitation.

Speleothems can be classified into three groups: circulating water deposits, oozing water deposits and stagnant water deposits. This classification takes into account the shapes and styles of the speleothems.

The deposits of circulating water represent the speleothems formed by the deposition of calcium carbonate contained in aqueous solutions that move through the caves due to gravity, such as stalactites, stalagmites, curtains, columns, travertine dams and cave pearls. These speleothems cover both the ceilings and walls of the cave, as well as its floor, and are formed by the two basic deposition mechanisms: dripping and oozing. This type of speleothem is the most commonly found in the caves studied.

Slalactites are the most commonly found speleothems. Drops of water containing calcium carbonate in solution, when they emerge from the cracks in the cave ceiling, remain stuck to them for a few minutes until they reach a volume heavy enough to overcome the surface tension and fall. During this time, carbon dioxide is released into the cave atmosphere, the solution becomes supersaturated and a delicate ring of calcite precipitates on contact between the drop and the ceiling. The stalactites grow vertically from the ceiling to the floor of the cave (figure 4.6).

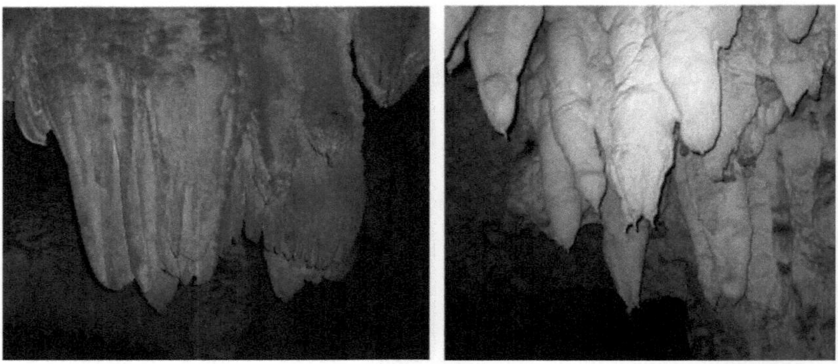

Figura 4.6: Stalactites from the Detrâs and Minotaur Caves, respectively.

By Barbara N. Rocha, 2009.

Stalagmites are formed when drops falling from the ceiling or from a stalactite hit the floor of the cave, precipitating the calcite they were still dissolved in. They grow vertically from the ground (figure 4.7).

Figura 4.7: Stalagmites from Jane Mansfield Cave.

By Mariana Rocha, March 2009.

The curtains are formed when water droplets emerge on a sloping wall or ceiling, run down its surface, leaving a thin trail of CaCO3 which, as the process continues, grows vertically, giving rise to a wavy, white and translucent calcite sheet (figure 4.8).

Figure 4.8: Curtains in the Cave of the Minotaur.

By Silvia Zambuzi, 2007.

Columns are the vertical and usually cylindrical shapes that originate from the joining of stalactites and stalagmites or from the overgrowth of one of them, joining the ceiling and floor of cave galleries and halls.

Figure 4.9: Column from the Minotaur Cave.

By Bàrbara N. Rocha, 2009.

Exudation water deposits are speleothems formed from aqueous solutions which, by capillarity, circulate slowly and discontinuously through the pores of the rock surrounding the cave. Various factors, such as the difference in temperature and pressure between the pores of the rock and the cavern void, cause these solutions to emerge from the walls, depositing the calcite that was dissolved until then. The speleothems representing this group are: corals, shields, helictites, cave flowers, needles, among others. Of this group of speleothems, only helectites could be observed in the caves surveyed.

Finally, deposits of stagnant water are speleothems that originate from the deposition of minerals in the submerged or superficial parts of water dams on the floors of caves, such as calcite geodes, which are not found in the Park.

In the next chapter (5), the results of the microclimatic studies carried out in the caves presented are presented and discussed.

5. Results and discussion

5.1 - Detrâs Cave (Cipó)

The Cipó Cave is small and consists of a single gallery. It is a cavity without running water, but with a high degree of dripping. There are some skylights in the central part, allowing for greater ventilation and the entry of diffuse solar radiation.

Field work to collect data in the cave took place between February 5 and 13, 2009. Figure 5.1.1 shows the locations where the thermohygrometers were installed.

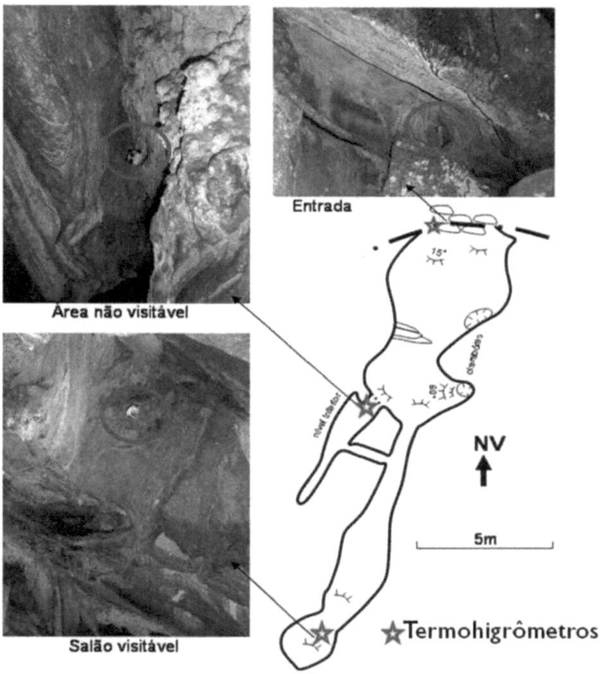

Figure 5.1.1: Location of the thermohygrometers in Detras Cave.

Organization and photographs: Barbara N. Rocha, 2009. Source: EKOS Institute, 2009.

The first was installed at the entrance to the cave, in a crack in the rock to protect it from the rain. The presence of ombrophilous forest in the external environment means that only diffuse solar radiation falls on this spot. This point characterizes the transitional microclimate, between the external and underground environments.

The second sensor, known as the "visitable room", was installed at the end of the gallery, in a narrower and more enclosed room, where the entry of light and atmospheric circulation was hampered by the narrow corridor and the absence of skylights.

As it consists of a single gallery, there are practically no areas in the cave where tourists don't go, which made it difficult to choose a place to install the third thermohygrometer (peripheral to a visitable area). So the last piece of equipment was placed in a small entrance in the middle of the gallery. This recess is small and very close to the circulation corridor. However, it is the only point in the cave where there is no passage for tourists.

Table 5.1.1 shows some of the statistical parameters (mean, maximum and minimum values, range, mode, variance and standard deviation) of the temperature and relative humidity data obtained at the three analysis points over a one-week period.

Table 5.1.1: Statistical parameters for temperature and relative humidity in Gruta do Cipó (period from 5 to 13/02/2009).

	Air temperature (°C)			Relative humidity (%)		
	Entry	Visitable room	Room not visitable	Entry	Visitable room	Room not visitable
average	19,5	17,8	18,9	100,0	100,0	99,2
maximum	22,7	18,5	19,9	100,0	100,0	100,0
minimum	16,7	16,4	18,0	100,0	100,0	70,8
range	6,0	2,1	1,9	0,0	0,0	29,2
fashion	19,4	18,3	18,8	100,0	100,0	100,0
standard deviation	1,2	0,8	0,4	0,0	0,0	3,8
variance	1,5	0,6	0,2	0,0	0,0	14,4

Organization: Barbara N. Rocha, 2010.

The average temperature during the study period was 19.5°C at the cave entrance, 18.9°C in the middle of the gallery (non-visitable room) and 17.8°C at the end of the cave (visitable room). It can be seen that the further you go into the cave, the lower the temperature. The average temperature range between the external and internal environments was 1.7°C.

The maximum temperature recorded during the period was 22.7°C at the entrance to the cave, where diffuse solar radiation can enter. The minimum (16.4°C) occurred at the end of the gallery, where there is no light, making it difficult to heat the atmosphere. At the entrance, the weekly temperature range was 6°C, while in the internal environment it was much lower, close to 2°C.

The mode of the samples was very close to the mean. This indicates that there were not many environmental fluctuations during the period analyzed, which can be seen in the low variance and standard deviation values, since the lower the variance, the less the observed values deviated from the mean.

The relative humidity showed averages close to or equal to 100% at the three points studied. This proves the high humidity in the underground environment. The maximum recorded at all points was 100%, while the minimum was 70.8% in the non-visitable area.

The amplitude in the middle of the gallery was 29.2 and at the entrance and end of the corridor the RH did not vary, the air being saturated throughout the period. The variation at the last point is due to the incidence of diffuse solar radiation and greater ventilation due to the presence of skylights.

The mode of the samples was always 100%. Thus, in all the locations analyzed, the air was saturated most of the time.

The variance and standard deviation of the middle of the gallery are high, due to the large amplitudes found.

The following graph (figure 5.1.2) shows the variation in air temperature at the three points studied over a period of one week. A two-axis graph was chosen to better visualize the amplitudes of the internal environment (visitable room and non-visible room), since the data from the external environment (entrance) is more variable.

During this first week in the field, the external weather station had problems downloading the data. Therefore, there is no comparison of the microclimate of this cave with that of the external environment.

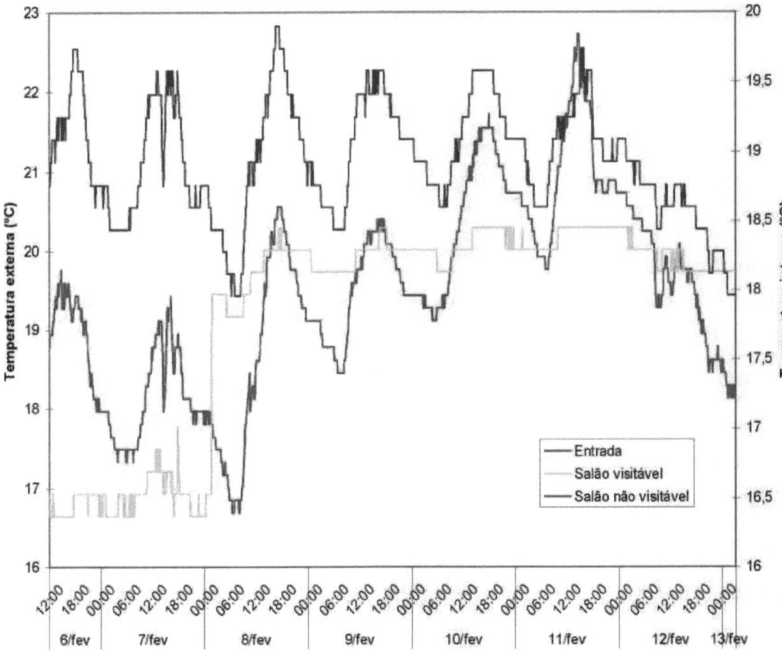

Figura 5.1.2: Air temperature variation (°C) in Gruta do Cipó.

Organization: Bàrbara N. Rocha, 2010.

It can be seen that the greatest thermal amplitudes are found at the entrance to the cave. The air temperature follows the diurnal cycle due to the greater incidence of solar radiation at this point.

As you enter the cave, the temperature range and air temperature gradually decrease. In the middle of the gallery (non-visitable area) the temperature still alternates between day and night, but with a smaller temperature range. There are hardly any delays in assimilating the weather outside. This is because this point is in a penumbra zone, with natural light from the cave entrance and skylights.

At the end of the gallery, the variations caused by the weather outside are still noticeable, but with much smaller amplitudes. The decrease or increase in temperature occurs almost simultaneously with the changes in the external environment. This indicates that this cave is not very susceptible to microclimatic changes caused by human visitation, as it has a high level of energy exchange and flow inside.

During the data collection period, a visit was made to the cave. This took place on the

7th, when a group of 5 people stayed in the cave from 5.30pm to 5.50pm. The variation in air temperature at the time of the visit can be seen more easily in the following figure (figure 5.1.3).

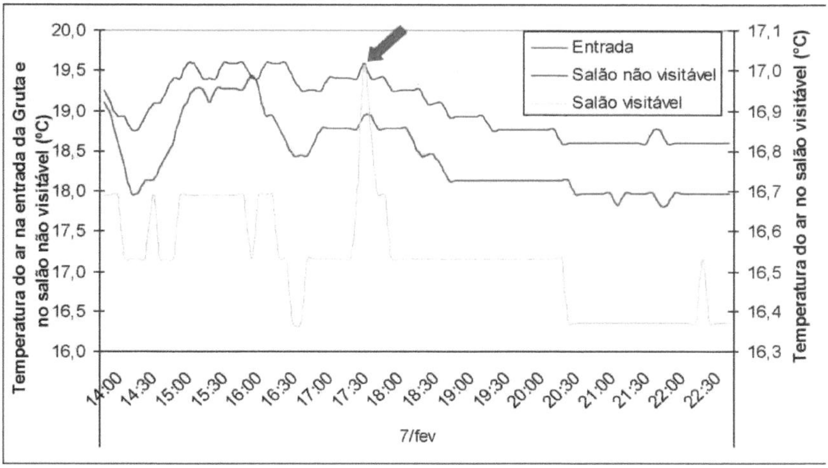

Figura 5.1.3: Hourly variation in air temperature (°C) in Gruta do Cipó on February 7th.

Organization: Barbara N. Rocha, 2010.

Figure 5.1.3 shows that, when the tourists entered, the air temperature at the end of the gallery rose by 0.4°C, from 16.5°C at 5.30pm to 16.9°C at 5.50pm. The cavity returned to a stable state at 6.05pm. This indicates an impact on the cave atmosphere due to tourist visits, corroborating the results found by Viana Jùnior (2002), Sànchez-Martos *et al* (2002), Longhitano *et al* (2006) and Rocha et *al* (2006).

The impact detected should be analyzed with caution, because at the time of the visit, the temperature of the external environment also showed an upward curve, increasing by 0.1°C. In the middle of the gallery, where variations still follow the diurnal cycle, the increase during this period was 0.2°C. A probable opening in the sky, with a reduction in cloudiness, may have facilitated the entry of diffuse radiation into the cave at the end of the day. Analysis of figure 5.1.2 shows that the visitable room responds to changes in the weather outside. Therefore, the increase in air temperature at the end of the gallery may, in part, have been due to external climatic variations.

As for the magnitude of the impact, a change of 0.4°C in 15 minutes at the innermost point of the cavity is not significant, as much larger variations, of a natural order, were found as early as dawn on the 8th, when the temperature of this same point rose by 1.5°C in a time interval of less than 15 minutes.

Figure 5.1.4 shows the variations in relative humidity in Gruta do Cipó.

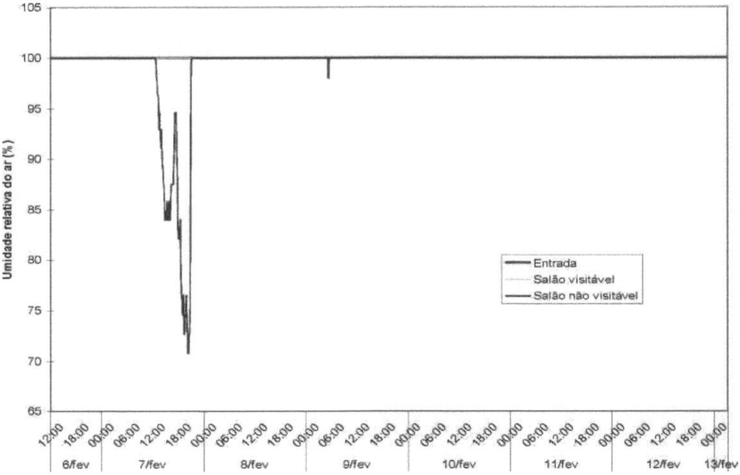

Figura 5.1.4: Variation in relative humidity (%) in Gruta do Cipó.

Organization: Barbara N. Rocha, 2010.

For most of the measurement period, the air was saturated at all three collection sites, including the cave entrance. This can be explained by the fact that the fieldwork was carried out during the rainy season and by the humidity coming from the forest outside the cave. In the underground environment, humidity tends to saturate due to the low incidence of solar radiation, the protection of the environment by the rock cover and the dripping of speleothems.

An analysis of the graph (figure 5.1.4) shows that on February 7, the sensor installed in the middle of the gallery (non-visible area) showed a variation in relative humidity, which reached a minimum of 70.8%. Two factors can explain this phenomenon: 1) the location of this point near the skylights, which allows more solar radiation to enter the area and consequently increases the temperature and decreases the RH (figure 5.1.2); 2) the presence of five tourists in the cavity, whose lighting and body heat can influence the microclimate of the underground environment.

It should be noted that the decrease in relative humidity at this location could already be seen at the start of the day, from 9.40am, reaching a minimum at 5.10pm and returning to saturation at 6.10pm (figure 5.1.5). The tourists didn't enter the cave until 5:30 p.m., when the atmosphere showed an increase in relative humidity.

Therefore, the first hypothesis is more coherent and the change in relative humidity due to tourism cannot be proven.

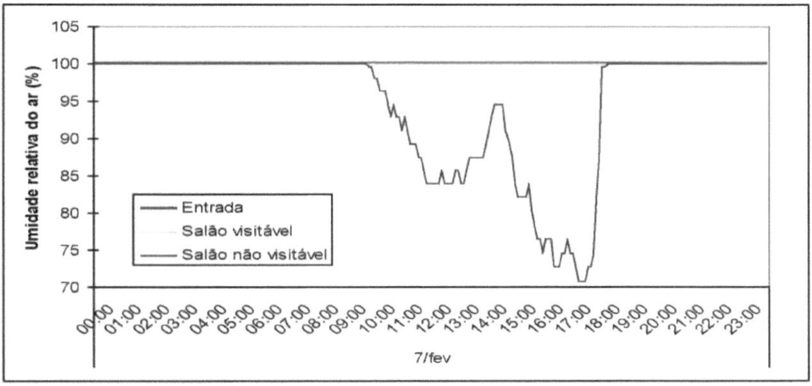

Figura 5.1.5: Hourly variation in relative humidity (%) at Gruta do Cipó on February 7, 2009.

Organization: Barbara N. Rocha, 2010.

The following figure (figure 5.1.6) shows the carbon dioxide profile of the cavity studied in the presence of three tourists. The data was collected on the morning of February 13th. The sketch of the cave shows the 10 data collection points, the first at the entrance and the tenth at the innermost point of the cave.

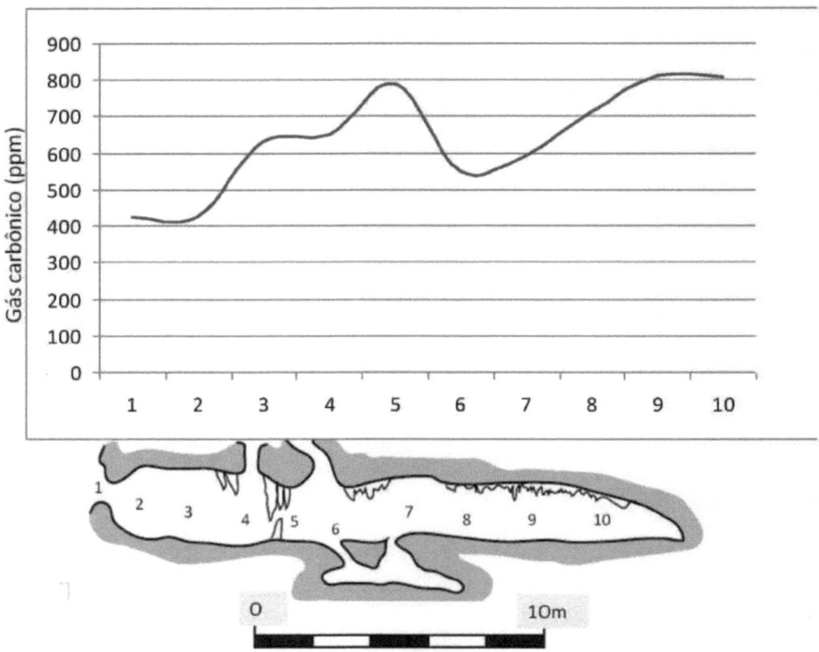

Figura 5.1.6: Atmospheric carbon dioxide profile of the Cipó Cave.

Organization: Barbara N. Rocha, 2010.

The amount of atmospheric carbon dioxide increases as you enter the cave, reaching a maximum of 810 ppm at the end of the gallery, compared to 425 ppm at the entrance. This upward trend is broken in the middle of the gallery due to the presence of skylights which allow for greater ventilation at this point.

The high concentration of CO_2 in the underground environment can be explained by the dynamic formation of speleothems, whose chemical process results in the release of gas. Other factors, such as the physiology of the cave, the production of oxidizing organisms and organic materials present in the substrate also contribute to the increase in the concentration of carbon dioxide, as described by Palomba (1995).

5.2 - Cave of Fire

The Gruta do Fogo is less than 130m long and consists of a dry part and a wet part, through which flows a narrow river with calm waters. It has no skylights and its interior is completely aphotic.

The temperature and relative humidity monitoring points can be seen in figure 5.2.1.

The first thermohygrometer (top left) was installed at the entrance to the cave, characterizing the transitional microclimate. At this point, diffuse solar radiation falls due to the dense vegetation cover outside. The second was placed in the middle of a dry corridor where people circulate. The third was at the end of the dry gallery, near the access to the river gallery, but outside the tourist trail.

Problems with the weather station prevented data from the cave from being compared with the outside environment.

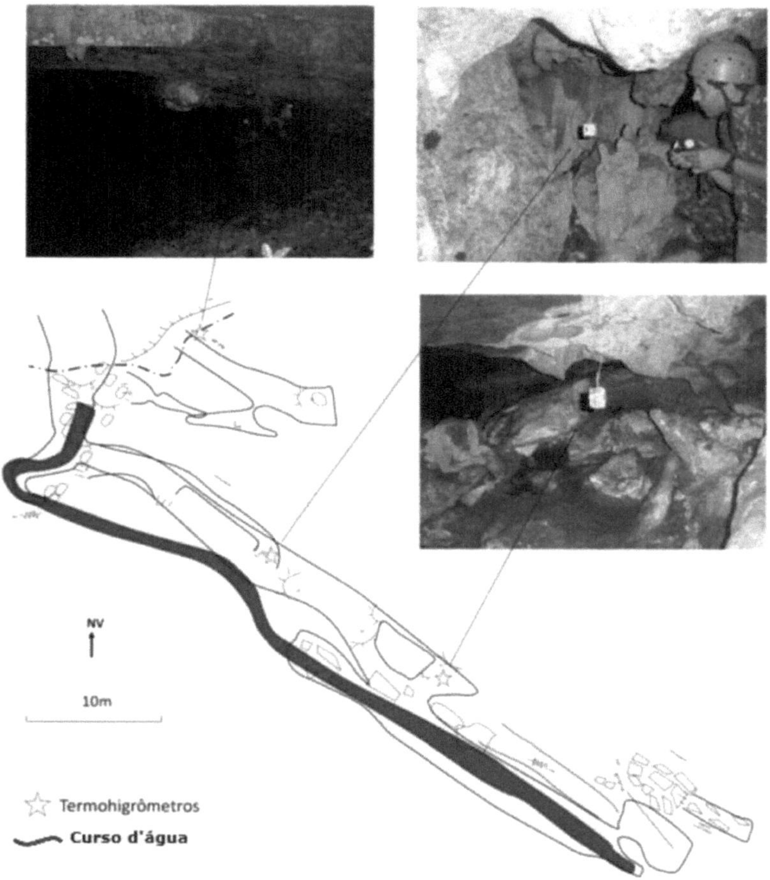

Figure 5.2.1: Location of the thermohygrometers in the Cave of Fire.

Organization and photographs: Bàrbara N. Rocha, 2010. Source: Instituto EKOS, 2009.

The following table shows some statistical parameters of the temperature and relative humidity data obtained at the three collection points from February 6 to 13, 2009.

Table 5.2.1: Statistical parameters for temperature and relative humidity in the Gruta do Fogo (period from 6 to 13/02/2009).

	Air temperature (°C)			Relative humidity (%)		
	Entry	Visible hall	Non-visitable hall	Entry	Visible hall	Non-visitable hall
average	19,1	17,2	16,8	100,0	100,0	99,3
maximum	20,4	17,5	17,1	100,0	100,0	100,0
minimum	17,3	17,2	16,8	96,1	100,0	70,8
range	3,1	0,3	0,3	3,9	0,0	29,2
fashion	19,3	17,2	16,8	100,0	100,0	100,0
standard deviation	0,6	0,0	0,1	0,2	0,0	3,7
variance	0,4	0,0	0,0	0,0	0,0	13,5

Organization: Bàrbara N. Rocha, 2010.

The highest average temperature for the period (19.1°C) was at the entrance to the cave, where there is direct and diffuse solar radiation. As you enter the cave, the average temperatures decrease, reaching their lowest value in the non-visitable area (16.8°C).

The highest value was recorded at the entrance to the cave, at 20.4°C. The lowest values were found in the internal environment, with 16.8°C in the non-visitable room and 17.2°C in the visitable room. The temperature range in the external environment was over 3°C and in the internal environment it was 0.3°C.

The modes of the temperature samples were equal to the average inside the cave. This stability of the environment can also be seen in the low variance and standard deviation.

The average relative humidity was 100% at the entrance and in the visitable room, and over 99% at the last point. The maximum recorded was always 100%, while the minimum was 70.8% in the non-visitable room. This high amplitude at this point contributed to a significant variance in the sample (14.3) and, consequently, to a high standard deviation.

The mode of the samples resulted in values of 100% at all points, confirming the predominance of situations with atmospheric saturation in cave environments.

The temperature and relative humidity statistics for the Fogo cave were similar to those for the Cipó cave. The relative humidity was high and the temperature showed smaller amplitudes as you went deeper into the cavities.

Figure 5.2.2 shows the variation in air temperature at the three data collection sites over a period of one week.

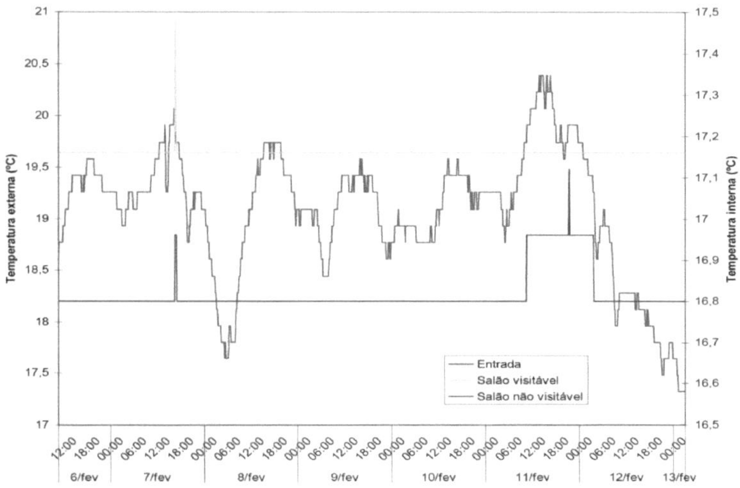

Figura 5.2.2: Air temperature variation (°C) in the Gruta do Fogo.

Organization: Bàrbara N. Rocha, 2010.

At the entrance to the cave, the air temperature showed a high amplitude, following the environmental fluctuations of the external environment (alternating between day and night). Inside the cave, the temperature remained constant at 17.2°C in the visitable room and 16.8°C in the non-visited area for practically the entire time. There was only a slight variation on February 7th and 11th.

The change on day 7 can be explained by the presence of five visitors, who were in the cave from 4.20pm to 4.40pm. If you look at the previous figure in more detail, you can see that the increase in temperature on that day occurred exactly when the visitors were in the cave (figure 5.2.3). The increase reached 0.3°C and 0.2°C in the visitable and non-visitable rooms, respectively. After the group left, the cave took around 20 minutes to return to its previous stable condition.

There were also variations in temperature on the 11th and 12th, only in the non-visitable room. This, being close to the riverbed, may have a greater flow of energy. The hypothesis is that the river could provide greater ventilation due to the flow of its waters, as well as having a different temperature to the atmosphere, promoting a variation in air temperature at this point.

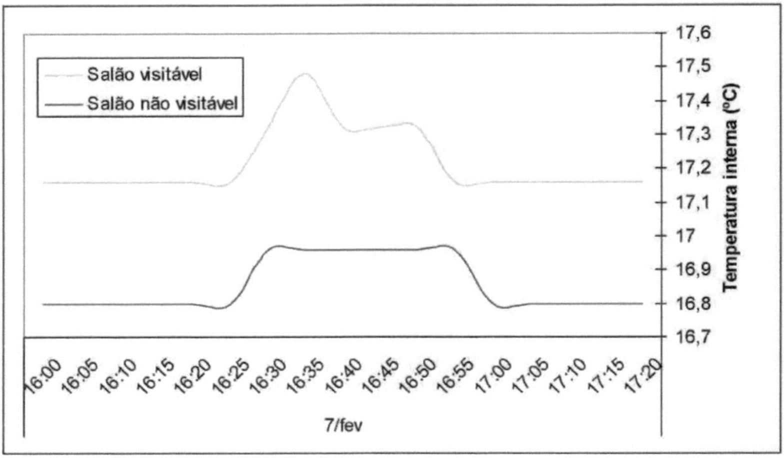

Figura 5.2.3: Air temperature variation (°C) inside the Cave of Fire every 5 minutes on February 7th.

Organization: Barbara N. Rocha, 2010.

Looking at the next figure (figure 5.2.4), the air was saturated for most of the period analyzed. The high humidity of the air is due to the protection provided by the rock cover, the presence of a watercourse inside the cave and the dripping of the speleothems.

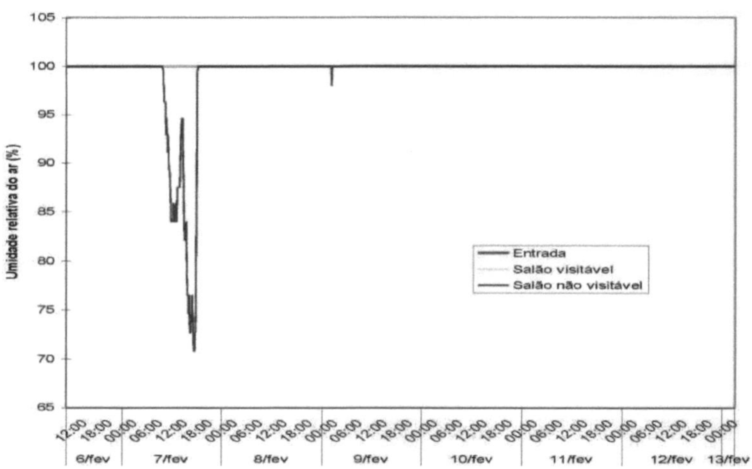

Figura 5.2.4: Variation in relative humidity (%) in the Gruta do Fogo.

Organization: Bàrbara N. Rocha, 2010.

At the entrance to the cave and in the visitable room, the relative humidity remained constant at 100% throughout the analysis period, while in the non-visitable room the RH varied greatly on day 7, reaching a minimum of 70.8%. This variation may be due to the greater flow of energy at this point, and may have been accentuated by the visitation, since the minimum was recorded at 5.10 p.m., shortly after the tourists left the cave.

The carbon dioxide profile of the Gruta do Fogo, as well as the monitoring points, is shown in figure 5.2.5.

CO_2 concentrations remain practically constant at 450 ppm in the cave entrance corridor, where ventilation is greatest, with a significant increase after the passage to the main gallery. In this second corridor, the rates were always close to 800 ppm. This increase is the result of the distance from the cave mouth, which hinders atmospheric circulation and the consequent dispersion of the gas. This proves that this cave is more susceptible to impacts on its microclimate from tourist visits.

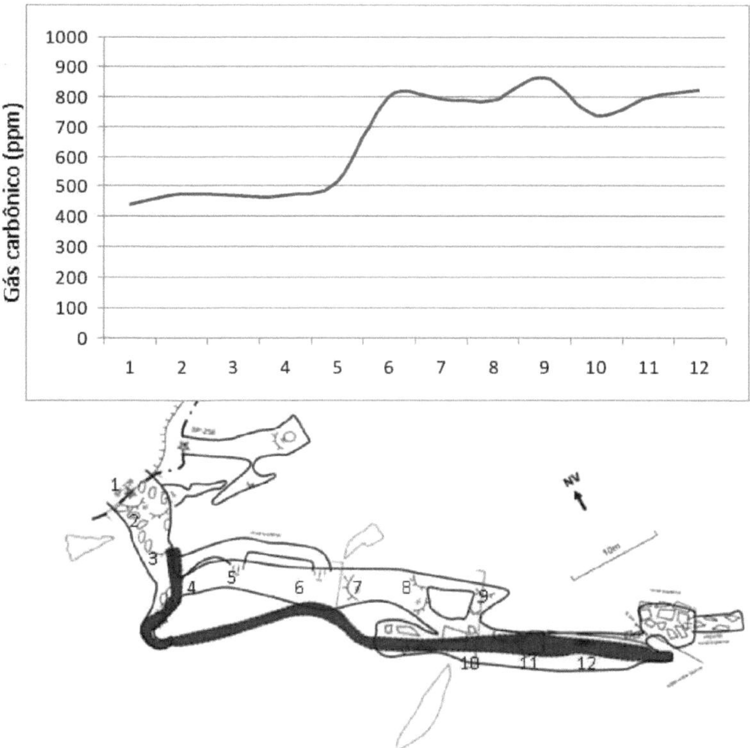

Figura 5.2.5: Atmospheric carbon dioxide profile of the Gruta do Fogo.

Organization: Barbara N. Rocha, 2010.

The analysis of the microclimate of the Gruta do Fogo showed that there are changes in the atmosphere of this cave due to tourist visits. It should be noted, however, that the changes in air temperature were close to the instrumental error of the thermohygrometers (0.2°C) for a maximum group of five tourists and that the cave recovers quickly thermally.

5.3 - The boys' playground

Toca dos Meninos is the smallest of the caves analyzed, measuring just 38m. It is a dry cave, meaning it has no watercourses and no aphotic zones, due to its small size and the presence of skylights. This configuration allows for good ventilation in the cavity, which directly influences its microclimate.

Figure 5.3.1 contains a map of the cave with the location and photographs of the sites chosen for monitoring.

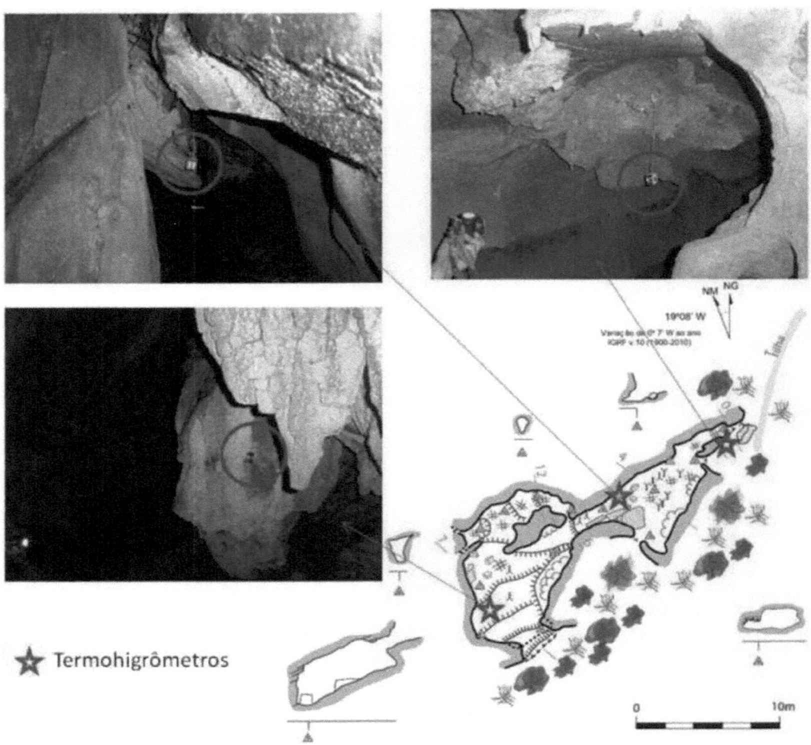

Figure 5.3.1: Location of the thermohygrometers at Toca dos Meninos.

Organization and photographs: Bàrbara N. Rocha, 2010. Source: EKOS Institute, 2009.

As defined in the methodology, the first thermohygrometer had to be installed at the entrance to the cave in order to characterize the transition microclimate between the external and internal environments. The second was installed in a place where tourists pass through, in the middle of the gallery, and the last was at the end of the corridor. Due to the configuration of this cavity, which is made up of a single small gallery, it does not contain any non-visitable sections.

The thermohygrometers were installed at Toca dos Meninos on the morning of February 6, 2009 and remained there until February 13. The data generated was statistically analyzed and can be seen in the table below.

Table 5.3.1: Statistical parameters for temperature and relative humidity at Toca dos Meninos (period from 6 to 13/02/2009).

	Air temperature (°C)			Relative humidity (%)		
	Entry	Visitable room	Room not visitable	Entry	Visitable room	Room not visitable
average	17,5	18,4	17,4	100,0	99,9	100,0
maximum	19,9	18,9	17,5	100,0	100,0	100,0
minimum	16,2	18,0	17,3	100,0	62,9	0,0
range	3,7	1,0	0,2	0,0	37,1	100,0
fashion	17,4	18,4	17,3	100,0	100,0	100,0
standard deviation	0,5	0,0	0,0	0,0	1,8	0,0
variance	0,7	0,2	0,1	0,0	1,3	0,0

Organization: Barbara N. Rocha, 2010.

The air temperature was higher in the middle of the gallery compared to the other analysis points, with an average of 18.4°C, compared to 17.5°C at the entrance and 17.4°C at the end of the corridor. The highest and lowest temperatures were recorded at the entrance (19.9°C and 16.2°C, respectively).

Thermal amplitudes were greatest at the entrance to the cave, with variations of 3.7°C. Inside the cavity, thermal variations during the period were equal to or less than 1°C.

The sample modes are very close to the averages. This fact, together with the low variances and standard deviations, suggests atmospheric stability in this cavity.

The relative humidity showed averages, maximums, minimums and modes equal to 100% at the entrance to the Toca and at the end of it. Consequently, the amplitudes, variances and standard deviations at these points were zero.

The RH only varied in the middle of the gallery; the minimum was 62.9%, giving a range of more than 37%. Despite the variability, the mode and median were 100%, proving that the air in the cave atmosphere is saturated most of the time. The variance was 1.8, resulting in a standard deviation of 1.3.

Figure 5.3.2 shows the variation in air temperature at Toca dos Meninos over the established period of one week. Due to problems with the weather station, there is no data from the outside environment to compare for this period.

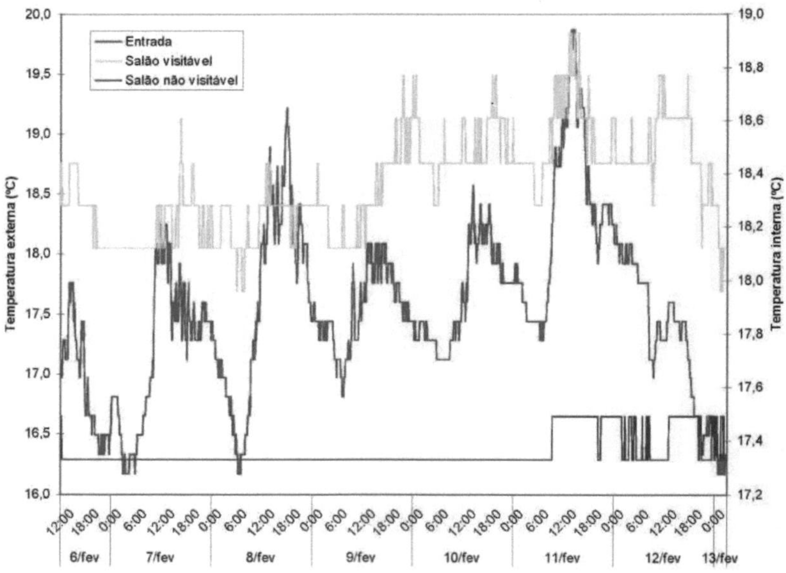

Figura 5.3.2: Air temperature variation (°C) at Toca dos Meninos.

Organization: Barbara N. Rocha, 2010.

The air temperature data at the point located at the entrance to the cave showed the highest daily oscillations, with a diurnal cycle that responds to the variation in solar radiation, since diffuse radiation is incident at this point. In the middle of the gallery (visitable room), the temperature was also variable, gently following the external variations, but with very small thermal amplitudes and a delay in the assimilation of the outside weather. At the end of the gallery (non-visitable room), the cave atmosphere tends towards stability, with a daily variation of less than 0.3°C.

Toca dos Meninos is rarely visited by tourists. Therefore, a visit had to be simulated on February 7 with five people to analyze the possible microclimatic changes in this environment.

In the presence of visitors, the temperature at the innermost point of the cave remained constant at 17.3°, a fact that demonstrates the absence of impacts on the cave's atmosphere from tourism.

The relative humidity can be seen in figure 5.3.3.

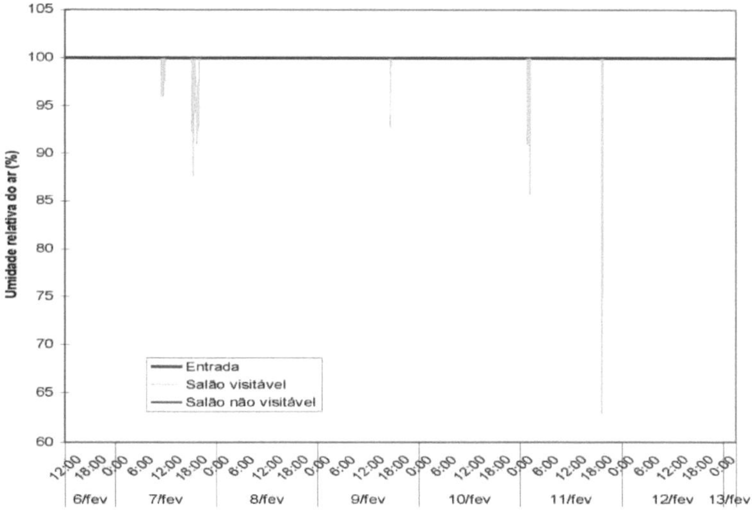

Figura 5.3.3: Variation in relative humidity (°C) at Toca dos Meninos.

Organization: Bàrbara N. Rocha, 2010.

The RH was constant at 100% at the entrance to the Toca dos Meninos and at the end of it. Variations can be seen in the middle of the gallery on three days during the period analyzed, with a minimum of 62.9% on 11/02. As the variations are frequent, we cannot relate them to tourism. These variations are due to the good ventilation of this point, located near the skylights and the entrance to the cave.

The saturation at the other two points is due to the dripping of the speleothems, the protection of the cave by the rock cover, the presence of closed forest in the external environment and the fact that the fieldwork was carried out during the rainy season.

The carbon dioxide profile of Toca dos Meninos was carried out on the morning of February 13th and can be seen below (figure 5.3.4). The profile was carried out in the presence of three people.

CO concentrations$_2$ ranged from 421 to 870 ppm, being lowest at the cave entrance and highest at the end of the corridor. There was a slight drop in concentrations at points near the skylights, due to increased ventilation.

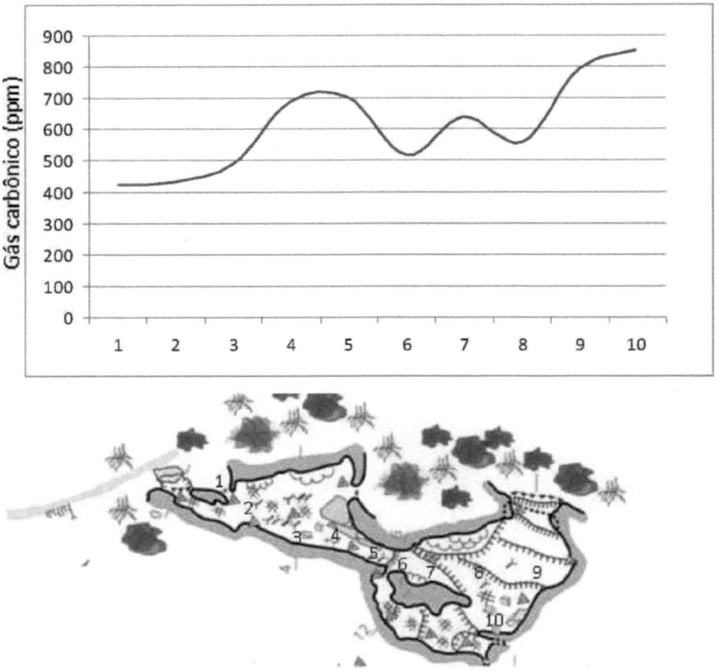

Figura 5.3.4: Carbon dioxide profile of Toca dos Meninos.

Organization: Barbara N. Rocha, 2010.

In short, the Toca dos Meninos has no changes in temperature or relative humidity as a result of tourist visits.

5.4 - Tatu Cave

The Tatu Cave is small (only 25m long). Its entrance is wide and from there you can follow two galleries, one to the front and one to the right. A narrow, shallow river flows into the gallery on the right, where there are also skylights to facilitate ventilation. From the front gallery it is possible to "see" the atmosphere outside, due to its proximity to the cave entrance. These characteristics create a great deal of energy exchange in the cavity.

The following photos show the microclimate monitoring points in Gruta do Tatu. The lack of a map of the cave made it impossible to draw a figure showing the location of the thermohygrometer installation points.

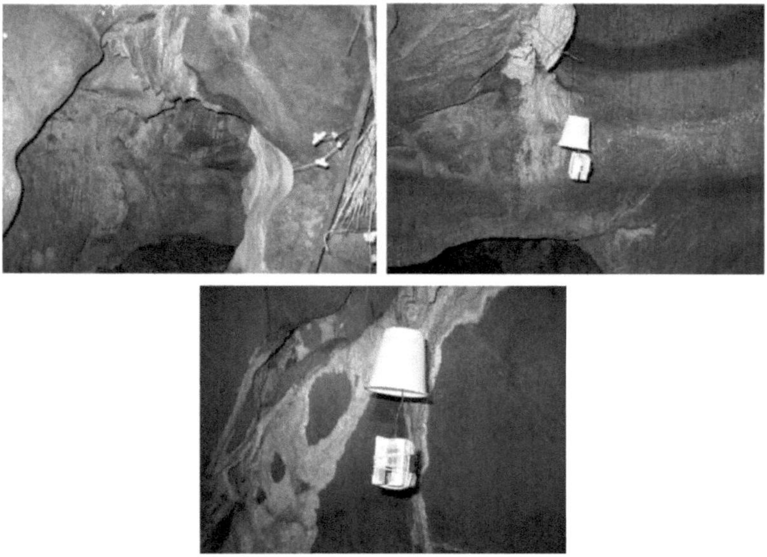

Figure 5.4.1: Photo of the thermohygrometers installed in the Tatu Cave.

Organization and photographs: Barbara N. Rocha, 2010.

The first thermohygrometer (top left) was installed at the entrance to the cave, characterizing the transitional microclimate. The second was installed at the end of the room on the right, where tourists head when they enter the cave, near the riverbed and the skylights (top right). The last one was in the room opposite the entrance, which is dry, i.e. not drained by the river, characterizing the area without visitors.

The temperature and humidity data were collected between February 28 and March 7. Their statistical analysis (mean, maximum, minimum, range, mode, median, variance and standard deviation) can be seen in the table below.

Table 5.4.1: Statistical parameters for temperature and relative humidity in Gruta do Tatu (period from 28/02 to 07/03/2009).

	Air temperature (^{0} C)			Relative humidity (%)		
	Entry	Salon Visitable	Room not Visitable	Entry	Salon Visitable	Room not Visitable
Average	20,8	19,6	19,7	100,0	100,0	100,0
Maximo	22,7	20,9	20,7	100,0	100,0	100,0
Minimum	19,1	17,6	18,0	100,0	100,0	100,0
Range	3,6	3,3	2,7	0,0	0,0	0,0
Fashion	20,7	19,4	20,1	100,0	100,0	100,0

Variance	0,7	0,4	0,3	0,0	0,0	0,0
Standard deviation	0,9	0,6	0,5	0,0	0,0	0,0

Organization: Barbara N. Rocha, 2010.

At the entrance to the cave, the average temperature was 20.8°C. Inside the cave, the averages were lower: 19.6°C in the visitable room and 19.7°C in the non-visited area. The maximum value recorded during the period was 22.7°C at the cave entrance and the minimum, 17.6°C, was recorded in the room with visitors.

The highest amplitude was recorded at the entrance to the cave, at 3.6°C. Inside, the amplitude was 3.3°C in the right-hand room and 2.7°C in the front room.

The mode of the samples was close to the mean at all points, indicating that the environment was thermally stable. The low variance and standard deviation values prove this.

The air was saturated at all the collection sites throughout the study period. Thus, the averages, maximums, minimums and modes were always 100%, while the amplitudes, variances and standard deviations were zero.

Figure 5.4.2 shows the variation in air temperature in Gruta do Tatu between February 28 and March 7. On the left-hand axis are the temperatures of the external environment (weather station) and the cave entrance, while on the right-hand axis you can see the temperatures of the internal environment (visitable and non-visitable rooms).

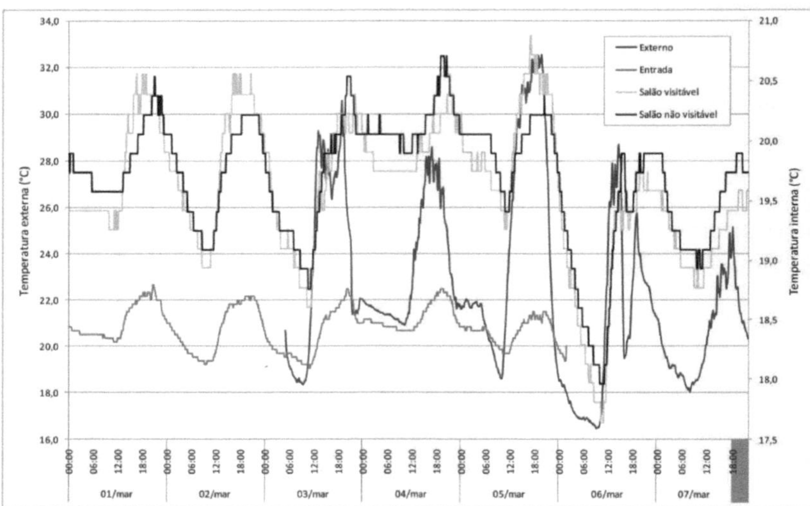

Figura 5.4.2: Air temperature variation (°C) in the Tatu Cave.

Organization: Barbara N. Rocha, 2010.

The thermograph installed at the entrance to the cavity had problems on March 5th, which made it impossible to record data from that day onwards. Despite this, it is possible to see that the variations in air temperature in the internal environment follow the external cycle, with a smaller thermal amplitude, and there are no delays in the assimilation of external atmospheric conditions. This cave, therefore, does not have a typical underground microclimate, which is characterized by stability.

Monitoring the diurnal temperature cycle in this cave is possible due to its configuration, with a wide mouth, skylights and a flowing river, which facilitate the exchange of energy with the external environment.

The Tatu Cave was visited by 10 tourists on March 7th. The first group, made up of 4 people, visited the cave from 9:15 am to 9:30 am. A second group, with 6 people, was in the cave from 5pm to 5.10pm. Despite the lack of records at the cave entrance on this day, it can be seen that the temperature in the internal environment showed the same variation, following the trend of external variations and without any peaks of change, which allows us to infer that there was no impact on the cave's atmosphere as a result of tourism. In addition, the air temperature dropped in the presence of the six tourists.

Figure 5.4.3 shows the variation in relative humidity in the Tatu Cave.

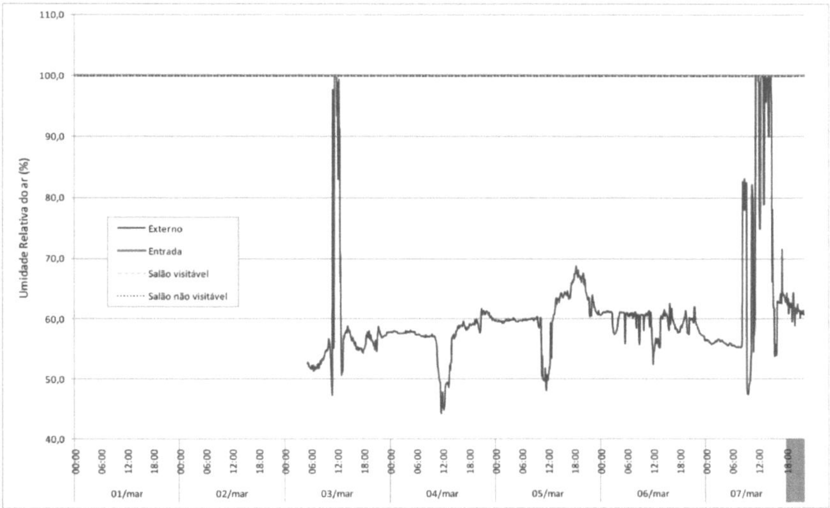

Figura 5.4.3: Variation in relative humidity (%) in the Tatu Cave.

Organization: Bàrbara N. Rocha, 2010.

Humidity at the entrance and inside the cave was constant at 100%, with no variations over the period. The cave's RH did not follow the external cycle, which fluctuated during the period due to the clear skies, as evidenced by the high direct solar radiation and lack of rainfall recorded by the weather station.

The high humidity in the cave's atmosphere is due to the protection by the rock cover, the evapotranspiration of the surrounding forest, the presence of a watercourse inside and the dripping of the speleothems.

Figure 5.4.4 shows the carbon dioxide profile taken on the morning of February 28, 2009. The lack of a map of the cave makes it difficult to see the points of the walk. The entrance to the cave is represented by the number 1 and the innermost point is marked with the number 12.

An analysis of the figure shows that the concentration of carbon dioxide rises as you enter the cave, decreases near the riverbed and rises again at the end of the gallery on the right.

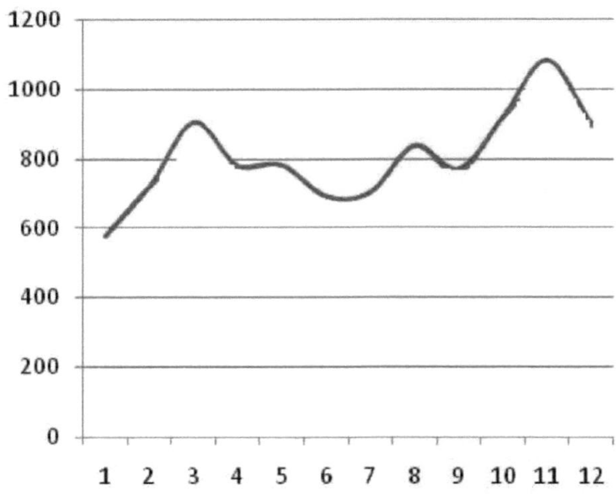

Figura 5.4.4: Atmospheric carbon dioxide profile of the Tatu Cave.

Organization: Barbara N. Rocha, 2010.

From the above, it can be concluded that there has been no change in the atmosphere of the Tatu Cave as a result of tourist visits. The presence of skylights, which promote greater ventilation and energy exchange with the external environment, seems to be a determining factor in the lack of microclimatic stability in this system.

5.5 - Jane Mansfield Cave

The Jane Mansfield Cave is approximately 500m long. It consists of a single corridor, with few bifurcations, which delimits a single visitation circuit. It is accessed via a small active spring from one of the tributaries of the Bocaina River, which runs through the entire cave. There are no skylights inside.

Field work to collect data at Jane Mansfield Cave took place between February 21 and 28, 2009. The photos below show the microclimate monitoring points in Jane Mansfield Cave (figure 5.5.1). The lack of a map of the cave prevented a figure from being drawn showing the location of the thermohygrometer installation points.

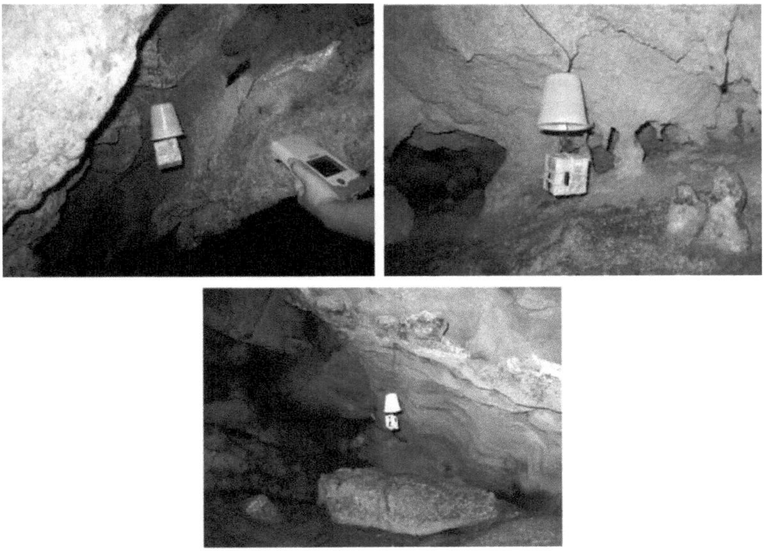

Figura 5.5.1: Photo of the thermohygrometers in Jane Mansfield Cave.

Organization and photographs: Barbara N. Rocha, 2010.

The first sensor (top left) was installed at the entrance to the cave, receiving diffuse solar radiation throughout the day. The second sensor (top right) was located at the end of the gallery (aphotic zone), where there are many visitors. The last thermohygrometer (bottom photo) was placed in an area outside the traditional visitor circuit, at the access to an upper gallery in the central part of the cave. At this point there is a water spout.

Table 5.5.1 shows the statistical parameters for temperature and relative humidity at the three sites studied over a one-week period. The parameters selected were: mean, maximum, minimum, range, mode, variance and standard deviation.

Table 5.5.1: Statistical parameters for temperature and relative humidity at Jane Mansfield Cave (period from 21 to 28/02/2009).

	Air temperature (°C)			Relative humidity (%)		
	Entry	Salon Visitable	Room not Visitable	Entry	Salon Visitable	Room not Visitable
Average	18,3	18,1	17,4	100,0	100,0	100,0
Maximo	19,8	18,3	17,6	100,0	100,0	100,0
Minimum	17,8	18,0	17,2	100,0	100,0	100,0
Range	2,0	0,3	0,4	0,0	0,0	0,0
Fashion	18,0	18,1	17,5	100,0	100,0	100,0
Variance	0,4	0,1	0,1	0,0	0,0	0,0
Standard deviation	0,2	0,0	0,0	0,0	0,0	0,0

Organization: Barbara N. Rocha, 2010.

Analyzing the temperature data, it can be seen that the weekly average was 18.3°C at the cave entrance, 18.1°C in the visitable room and 17.4° in the non-visited area. The higher temperature at the first point is explained by the incidence of diffuse solar radiation at this location, while the lower temperature at the last is due to the intense flow of water from the river, which increases ventilation at this point.

The maximum values for the period followed the averages, being highest at the entrance (19.8°C) and lowest in the non-visitable room (17.6°C). The period's minimum of 17.2°C was detected here. There was little variation between the highs and lows, with the temperature range being equal to or less than 2°C in all the locations surveyed. The small range is due to the high humidity of the cave, which is crossed by a river of running water throughout its interior. Water is a fundamental element in maintaining heat.

The modes are close to the averages at all points, confirming the cave's thermal stability.

The variability of the samples was small, resulting in a low standard deviation (equal to or less than 0.2 at all collection points). This is due to the low temperature range, which indicated great stability in the data.

Throughout the analysis period and at all the recording points, the air was saturated. The mean, maximum, minimum and mode parameters were always 100% and the amplitude, variance and standard deviation of the relative humidity were zero.

Saturation is due, among other factors, to the presence of a flowing river with a few waterfalls, a constant source of moisture for the atmosphere.

The following graph (figure 5.5.2) shows the variation in air temperature at the three monitoring points in the cave and in the external environment.

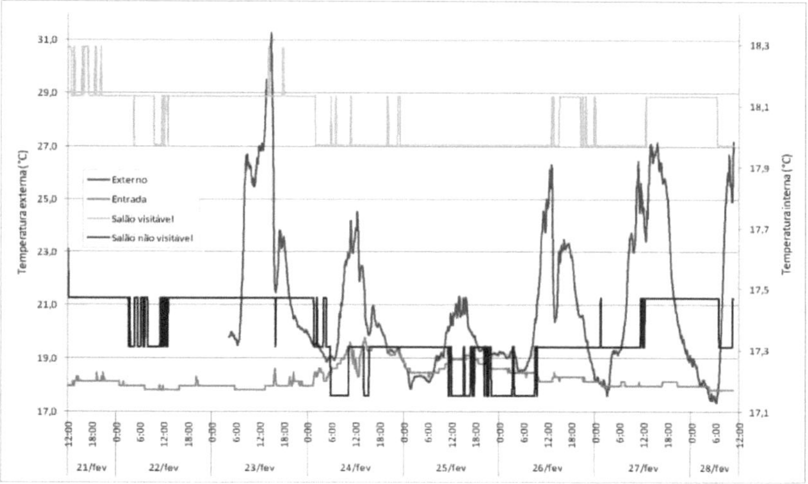

Figura 5.5.2: Air temperature variation (°C) in Jane Mansfield Cave.

Organization: Barbara N. Rocha, 2010.

At the entrance to the cave, the temperature varies in line with the diurnal cycle, but with reduced amplitudes and some delays in the assimilation of the weather outside. From February 26 to 28, the temperatures at this point did not keep up with the rises in the external environment. The presence of a river of running water at the entrance to the cave brings a lot of humidity into the atmosphere, preventing large variations. The biggest temperature fluctuations at this point occurred between February 24 and 25, reaching a maximum of 19.8°C.

In the visitable area, temperatures remained practically constant at 18°C. The non-visitable area had the lowest temperatures, due to its location close to the riverbed. The lowest temperature, 17.2°C, was recorded on February 24, 25 and 26.

During the recording period, the cave was visited by five groups of tourists. On the 22nd, a group of three visited the cave from 10:00 am to 10:35 am and another, with six members, from 10:45 am to 11:45 am. On the morning of that day, the two internal recorders showed the lowest temperatures of the day.

A group of six tourists visited the cave from 1pm to 2.20pm on the 23rd. No change in temperature was detected in the non-visitable room. The visitable room, on the other hand, recorded an increase of 0.1°C, which is within the instrumental error of 0.2°C, over a ten-minute interval, from 2pm to 2.10pm. Changes of this magnitude also occurred on days and periods without visitors, which does not prove the occurrence of an anthropogenic impact.

Another eight people visited the cave on the mornings of the 24th and 27th. At the time of the visits, no changes were found in the temperature curves.

Figure 5.5.3 shows the variation in relative humidity. The humidity in the cave did not vary during the period, being 100% at all points, thanks to the constant presence of water in the gallery, due to the presence of a river, the dripping of speleothems and the protection of the rock cover.

Even at the entrance to the cave, variations in the relative humidity of the air outside could not be perceived.

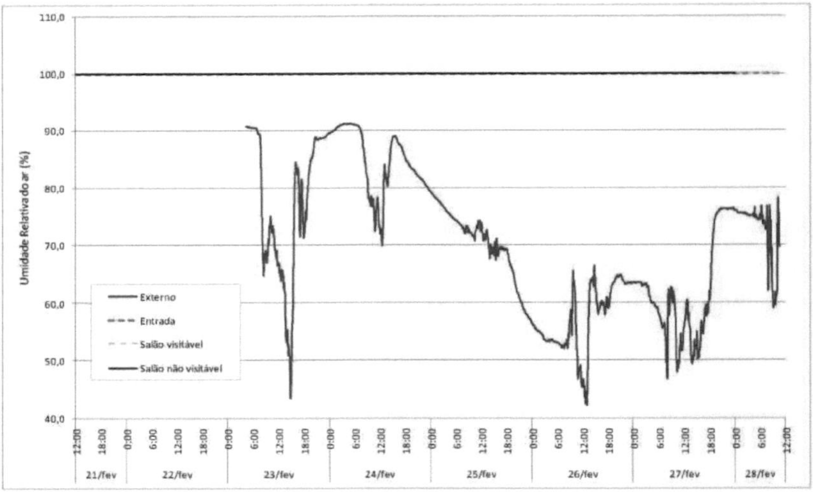

Figura 5.5.3: Variation in relative humidity (%) at Jane Mansfield Cave.

Organization: Bàrbara N. Rocha, 2010.

Figure 5.5.4 shows the carbon dioxide profile of Jane Mansfield Cave, taken on the morning of February 28. Data was collected at 15 points in the cave, from the entrance (point 1) to the end of the gallery (point 15).

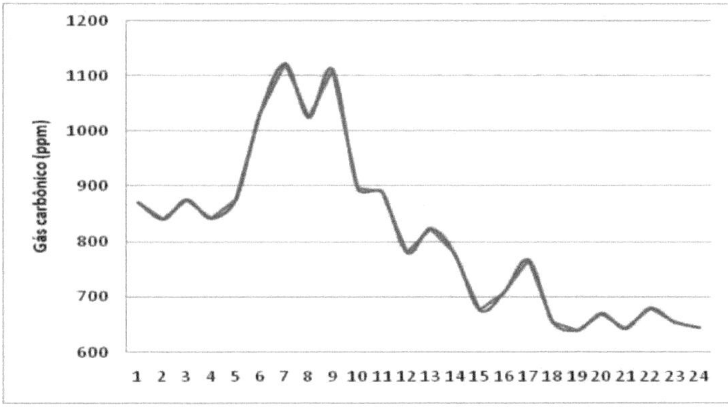

Figure 5.5.6: Atmospheric carbon dioxide profile of Jane Mansfield Cave.

Organization: Bàrbara N. Rocha, 2010.

At the entrance to the cave (point 1) the concentration of carbon dioxide was 870 ppm. After passing through a narrow corridor that gives access to the cave, the CO_2 level rises significantly, reaching 1122 ppm in the first room (point 7). The gas concentration then slowly decreases until it reaches 645 ppm at the end of the cavity (point 24). Empirically, greater ventilation can be seen from point 7 onwards, as the river becomes more flowing and has small waterfalls.

The methodology adopted concludes that the Jane Mansfield Cave does not show changes in its atmosphere (temperature, relative humidity and carbon dioxide) during periods of tourist visitation.

5.6 - Fendao Cave

The Fendao Cave is one of the largest in the Intervales State Park, stretching for over a kilometer. It is a wet cave, formed by the river gallery that crosses its entire interior. It has two entrances and several skylights, as well as high ceilings and large halls, allowing for a great exchange of energy with the external environment. The intense flow of water from the river, combined with its large volume, also facilitates this exchange.

All hiking is done on the riverbed. In times of flood, a tourist of average altitude can wet their body up to chest height.

Figure 5.6.1 shows a map of the cave, photos of the monitoring points and where the thermohygrometers were installed.

Fieldwork to collect data from the Fendao Cave took place between February 28 and

March 7, 2009. The first thermohygrometer (photo on the right) was installed at one of the cave entrances, at the river resurgence. This location has a rocky wall that protects it from the sun's light, allowing only diffuse solar radiation to pass through at certain times of the day.

The second recorder was placed in the middle of the cavity, in a larger room where tourists stop for a *blackout*. In addition to the riverbed, this spot has a dry section. Near this point, there is an entrance that gives access to a small upper gallery. This spot was chosen for the installation of the third thermohygrometer (top left).

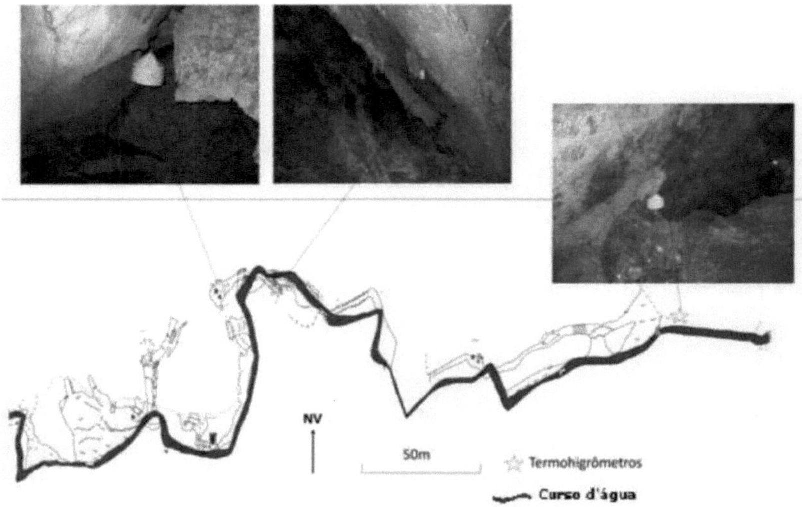

Figure 5.6.1: Location of the thermohygrometers in the Fendao Cave.

Organization and photographs: Barbara N. Rocha, 2010. Source: EKOS Institute, 2009.

The resulting data was analyzed using statistical parameters and can be seen in table 5.6.1.

The thermograph located at the entrance to the cave had problems and only started recording in the early hours of March 3rd. Considering this shorter period, the average recorded at this location was 15.8°C, lower than the internal averages of 18.0°C in the visitable room and in the non-visited area. As mentioned above, the entrance to the cave has a rocky wall that protects it from the sun, allowing the temperature to drop. Inside, the water from the river, the high humidity and the rock cover prevent the temperature from dropping during the early hours of the morning.

Table 5.6.1: Statistical parameters for temperature and relative humidity in the Fendao

Cave (period from 28/02 to 07/03/2009).

	Air temperature (° C)			Relative humidity (%)		
	Entry	Salon Visible	Room not visible	Entry	Salon Visible	Room not visible
Average	15,8	18,0	18,0	100,0	100,0	100,0
Maximum	16,5	18,1	18,1	100,0	100,0	100,0
Minimum	14,6	17,8	17,8	100,0	100,0	100,0
Range	1,9	0,3	0,3	0,0	0,0	0,0
Fashion	16,0	18,0	18,0	100,0	100,0	100,0
Variance	0,2	0,0	0,0	0,0	0,0	0,0
Standard deviation	0,4	0,1	0,1	0,0	0,0	0,0

Organization: Bàrbara N. Rocha, 2010.

The highest temperatures were found inside the cave, with 18.1°C in both rooms. The minimum occurred at the entrance, due to the configuration of this point, and was 14.6°C.

The thermal amplitudes in Gruta do Fendao were low: 1.9°C at the entrance and 0.3°C at the internal monitoring points. This atmospheric stability is due to the high humidity in the environment, which can be seen by analyzing the relative humidity data. At all the sites studied, the air was saturated throughout the week, with averages, maximums, minimums and modes equal to 100% and zero amplitudes, variances and standard deviations. The river is the main provider of humidity in this cave.

The mode of the indoor temperatures was equal to the mean, reinforcing the low environmental fluctuations of these sites. The variances tended towards zero and the standard deviation was 0.1. At the entrance, the most commonly found value was 16.0°C, higher than the average. The variance was 0.2 and the standard deviation was 0.4.

Figure 5.6.2 shows the variation in air temperature in Gruta do Fendao over the course of the week. The left axis shows the data from the external environment (weather station) and the cave entrance. The data from the visitable and non-visitable rooms can be seen on the right-hand axis.

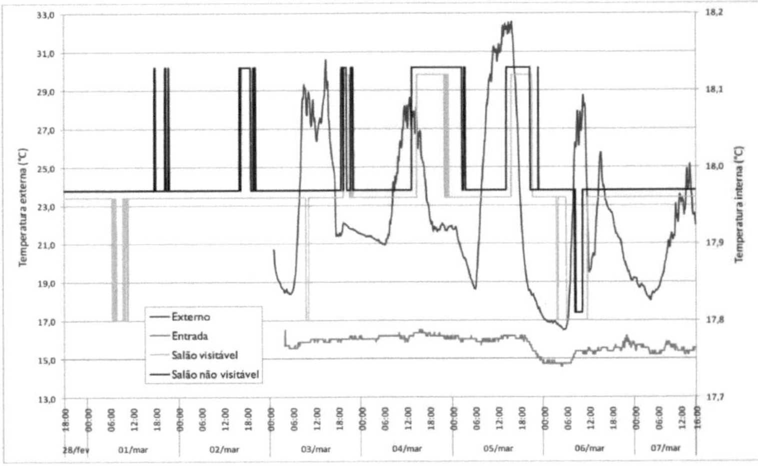

Figure 5.6.2: Air temperature variation (°C) in the Fendao Cave.

Organization: Bàrbara N. Rocha, 2010.

Despite the problems with the first thermohygrometer, which prevented records from being taken during the first few days, it can be seen that the microclimate at the entrance to the cave follows the trend of the external environment, with an increase in temperature during the day and a decrease at night. However, the amplitudes are quite small. Some variations in the external environment are not perceived at the cave entrance.

At the points located inside the cave, the air temperature showed small variations, in the order of a few tenths, which is the result of the sensor's accuracy. In the non-visitable area, at an access to the upper gallery, the air temperature was slightly higher than in the visitable room. In general, the two points showed the same trend of change.

The Fendao Cave is one of the most visited in the park and received nine tourists during the period, three who toured the cave on March 4 and six on March 7. On these days, the thermograph at the entrance was working again, allowing for a better analysis.

The group of three tourists entered the cave at 9:30 am and left at 10:30 am. During this time, the temperature outside remained practically constant at 16.1°C. Inside the cave, the temperature did not change at all. On the 7th, before the equipment was removed, six tourists toured the cave. Temperatures did not change in the presence of this group either.

The following graph (figure 5.6.3) shows the variation in relative humidity in the Fendao Cave during the study period.

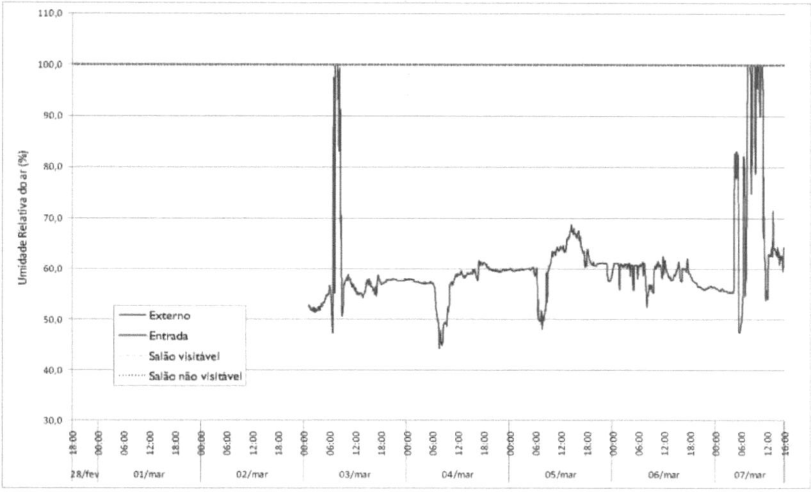

Figura 5.6.4: Variation in relative humidity (%) in the Fendao Cave.

Organization: Barbara N. Rocha, 2010.

The relative humidity remained at 100% at all three points in the cave throughout the period. Even the recorder at the entrance to the cave didn't keep up with the variations in the external environment. Saturation in this cave is due to the presence of the river with its high volume and flowing water and the rocky capping. At the entrance, the vegetation prevents direct solar radiation from entering and provides moisture to the atmosphere through evapotranspiration.

The carbon dioxide profile of the cavity is shown in the figure below. The CO_2 data was collected on March 7th, in the late afternoon.

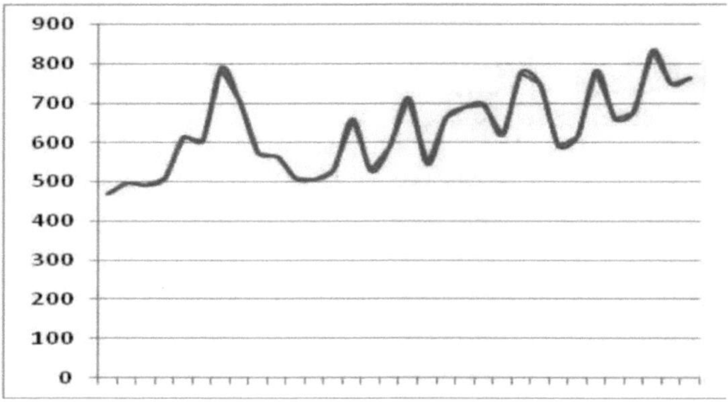

Figura 5.6.5: Carbon dioxide profile of the Fendao Cave.

Organization: Barbara N. Rocha, 2010.

The concentrations of carbon dioxide varied greatly, with no trend in concentration. What could be empirically perceived is that the gas concentration varies according to the ventilation of the cave and not by its configuration. As energy exchange is intense in this cave and ventilation is facilitated by the high ceiling, the presence of skylights and the flow of the river, the breezes that blew at times were responsible for the reduction in CO_2 concentration.

This cave has the most difficult route, forcing visitors to make a greater physical effort on their way through, which would release a greater amount of carbon dioxide into the interior. However, the good ventilation of the Fendao cave facilitates atmospheric exchange, preventing an increase in gas concentration.

From the above, it can be concluded that Gruta do Fendao did not show any changes in its atmosphere in the presence of tourists. This cave is unlikely to change with larger groups due to its high capacity for energy exchange with the external environment.

5.7 - Maozinha Cave

The Maozinha Cave belongs to the same complex as the Fendao Cave and is located near one of its entrances. It is a small cave, consisting of a single gallery. Its entrance overlooks a vegetated valley floor, through which flows the river that crosses the Fendao Cave. There are no bodies of water inside, but the soil and walls of the cavity are extremely wet and covered in clay. There are no skylights or secondary entrances, and the interior is completely aphotic.

Figure 5.7.1 shows the location of the temperature and relative humidity loggers in

Gruta da Maozinha. They were in the field from February 13 to 20, 2009.

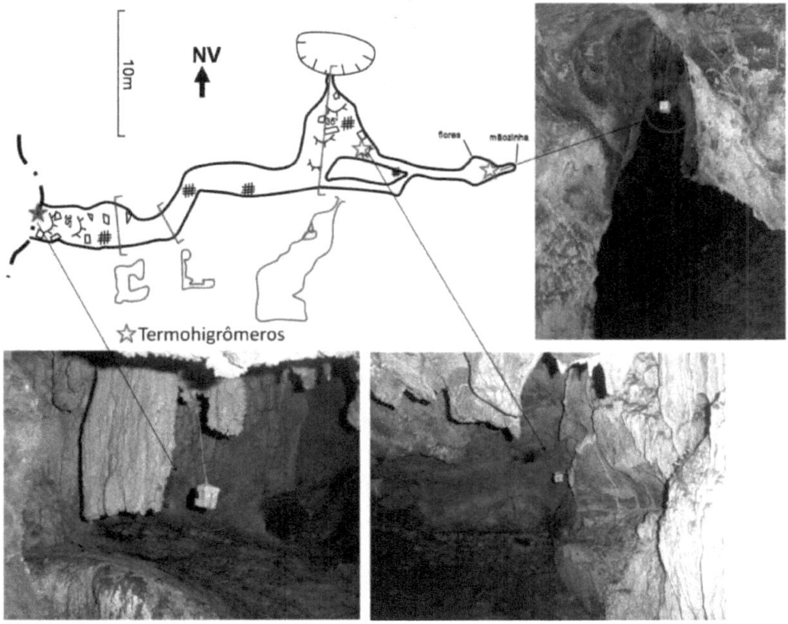

Figure 5.7.1: Location of the thermohygrometers in the Maozinha Cave.

Organization and photographs: Barbara N. Rocha. Source: EKOS Institute, 2009.

The first sensor was installed at the entrance to the cave in order to characterize the transition microclimate between the internal and external environments. The solar radiation that reaches this point is mainly diffuse, due to the presence of tree vegetation in front of the cave.

The second sensor was installed at a fork in the middle of the gallery. The preferred route to reach "Maozinha", the speleothem that gave the cave its name, is to the right. So the sensor was installed in the passage on the left, the only area of the cave that is not usually visited by tourists.

The third thermohygrometer was installed at the end of the gallery, near the "little hand". This is the most visited spot in the cave.

The data generated at the three points was processed statistically and can be seen in the table below. The averages, maximums, minimums, ranges, modes, medians, variances and standard deviations of the temperature and humidity samples were extracted.

Table 5.7.1: Statistical parameters for temperature and relative humidity of the air in Gruta da Maozinha (period from 13 to 20/02/2009).

	Air temperature (^{0} C)			Relative humidity (%)	
	Entry	Salon Visitable	Room not Visitable	Salon Visitable	Room not Visitable
Average	18,5	17,0	17,7	100,0	100,0
Maximo	21,7	17,8	18,5	100,0	100,0
Minimum	17,2	16,7	17,2	100,0	100,0
Range	4,5	1,1	1,3	0,0	0,0
Fashion	17,2	17,0	17,5	100,0	100,0
Variance	1,0	0,0	0,1	0,0	0,0
Standard deviation	1,0	0,2	0,3	0,0	0,0

Organization: Barbara N. Rocha, 2010.

The highest average temperature was found at the entrance to the cavity, at 18.5°C. In the middle of the gallery the average was 17.7°C and at the end 17.0°C. The thermohygrometer located at the entrance also showed the highest temperature for the period (21.7°C). The lowest temperature was in the visitors' lounge (16.7°C).

As expected, the greatest amplitude, 4.5°C, occurred at the entrance to the cave, where the incidence of diffuse solar radiation causes the temperature to rise during the day and the absence of protection by the rock cover allows it to drop more sharply at night. At the points located in the internal environment, the amplitudes did not exceed 1.3°C.

The mode calculated at the entrance was 17.2°C. Inside the cave, the modes were close to the average.

Due to the greater variations, the entrance had the greatest variance and standard deviation (1.0). In the visitable and non-visitable halls, these parameters were lower, with the standard deviation of the samples being 0.3 in the middle of the gallery and 0.2 at the end of it.

The hygrometer located at the entrance to the cave had problems due to water entering the sensor as a result of runoff from the rock protecting it. As a result, data on this attribute could not be collected at this point. The same problem also occurred with the thermograph at this point, which stopped working on February 18th at 2:30pm.

Indoors, the air was saturated throughout the period. Therefore, the averages, maxima, minima and modes were equal to 100%. The amplitudes, variances and standard deviations were therefore zero.

Figure 5.7.2 shows the variation in air temperature in Gruta da Maozinha and compares it to the data obtained from the external weather station. On the left-hand axis, you can see the data from the outside and the entrance to the cave. The data from inside the cavity is on the right.

At the entrance to the cave, the temperature followed the diurnal cycle, since from this point you can "see" the outside atmosphere, with direct and diffuse solar radiation coming in. Over the course of the week, the temperature rose during the day, increasing the daily temperature range. The points located indoors followed this cycle, but with smaller amplitudes and a delay in assimilating the weather outside.

The small size of the cave and its location in front of a valley floor means that energy is exchanged in all the rooms. As a result, the temperature does not remain constant, but always fluctuates.

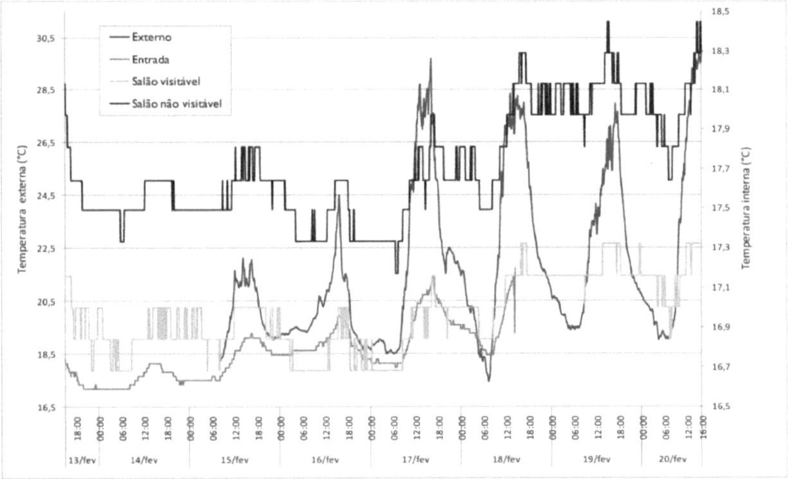

Figura 5.7.2: Air temperature variation (°C) in the Maozinha Cave.

Organization: Bàrbara N. Rocha, 2010.

On February 15, a group of three tourists visited the cave, staying inside from 12:25 pm to 12:35 pm. On this day, the temperature at the innermost point of the cave had been rising since 11.20 am, and remained constant at 17.6°C until 2.30 pm. It can therefore be concluded that the group of tourists had no impact on the cave atmosphere.

The relative humidity data can be seen in figure 5.7.3.

As previously analyzed from the data shown in table 5.7.1, the relative humidity did not vary inside the cave, remaining constant at 100%. Air saturation could be seen in the field by the constant humidity of the rock walls and the wet, very smooth soil.

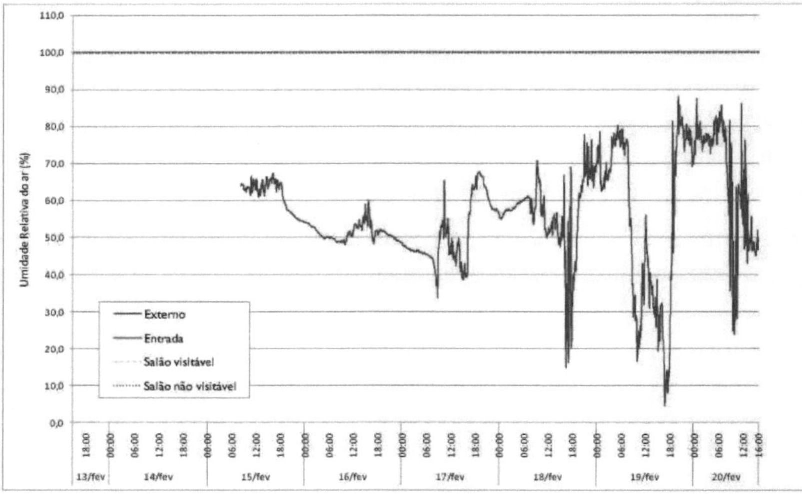

Figura 5.7.3: Variation in relative humidity (°C) in the Maozinha Cave.

Organization: Bàrbara N. Rocha, 2010.

The next figure shows the carbon dioxide profile of the Maozinha Cave (figure 5.7.4). The profile was carried out on February 20th in the presence of four people.

Under natural conditions, carbon dioxide ranges from 455 ppm to 760 ppm. At the entrance, the CO_2 level was lower, rising in the central part of the gallery and falling again in the "little hand" room.

We decided to carry out a visitor test using a carburetor. After the device was switched on, the concentration of carbon dioxide at the end of the gallery jumped from 760 ppm to 1695 ppm, as can be seen in the figure. Even after the carburetor was switched off, CO_2 concentrations remained high until the group left. This result is in line with the work of Scaleante (2003), who proved that environments impacted by the presence of visitors using carbide lighting take time to return to the initial situation.

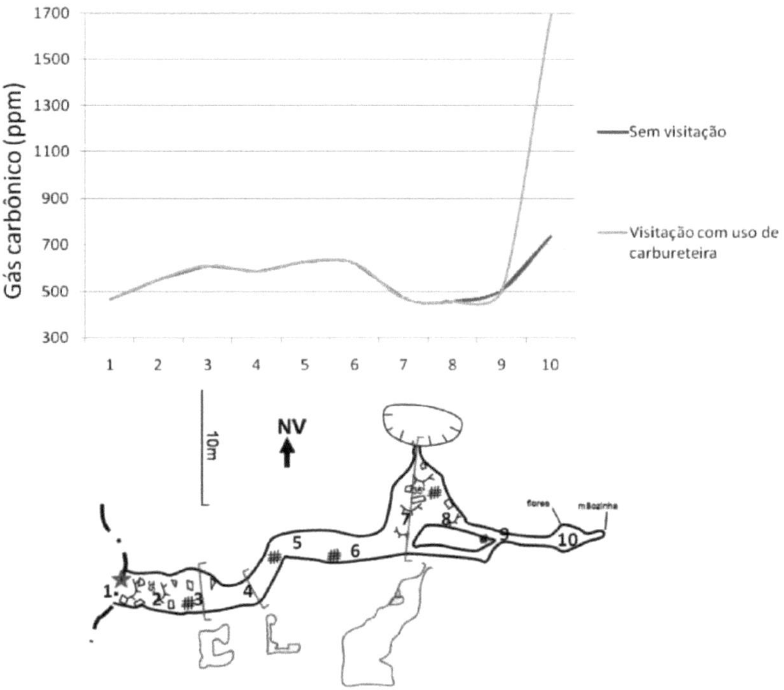

5.7.4: Carbon dioxide profile of the Maozinha Cave.

Organization: Barbara N. Rocha, 2010. Source: EKOS Institute, 2009.

As a result, no impacts were found on the atmosphere of the Maozinha Cave under normal visiting conditions (without the use of carburetors). However, it was possible to conclude that tourism with the use of carburetors results in a significant increase in carbon dioxide levels.

5.8 - Santa's Cave

The Grotto of the Saint has a single wide entrance which gives it good ventilation. It consists of a single gallery with wide corridors. A small watercourse runs through part of its interior.

The temperature and relative humidity monitoring points in Gruta da Santa can be seen in figure 5.8.1. The data was collected between February 14th and 20th.

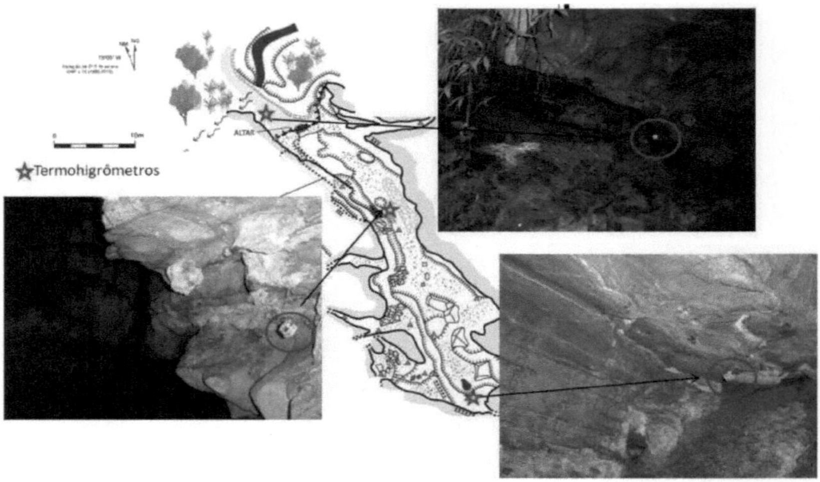

Figure 5.8.1: Location of the thermohygrometers in Gruta da Santa. Organization and photographs: Bàrbara N. Rocha, 2010. Source: EKOS Institute, 2009.

The first thermohygrometer was installed at the entrance to the cave, next to the altar. As the entrance to the cave is wide and the vegetation in front of it consists mainly of short grass, this location receives direct and diffuse solar radiation. The north-facing mouth allows the radiation to shine in throughout the day.

The next monitoring site is in the middle of the cavity, on the traditional visitor circuit, next to the watercourse and a travertine. Here, visitors stop to admire the speleothem, making it a popular spot for visitors.

The last thermohygrometer was installed at the end of the pipeline, where there are no tourist visitors.

Table 5.8.1 shows some of the statistical parameters calculated from the temperature and relative humidity data collected throughout the week. The parameters chosen for analysis were: mean, maximum, minimum, range, mode, variance and standard deviation.

Table 5.8.1: Statistical parameters for temperature and relative humidity in Gruta da Santa (period from 14 to 20/02/2009).

	Air temperature (^{0} C)			Relative humidity (%)		
	Entry	Salon Visitable	Room not Visitable	Entry	Salon Visitable	Room not Visitable
Average	16,7	17,2	17,1	99,9	100,0	100,0
Maximo	19,4	17,5	17,3	100,0	100,0	100,0

Minimum	14,1	17,0	16,7	68,4	100,0	100,0
Range	5,3	0,5	0,7	31,6	0,0	0,0
Fashion	16,8	17,3	17,2	100,0	100,0	100,0
Variance	1,4	0,0	0,0	2,8	0,0	0,0
Standard deviation	1,2	0,1	0,1	1,7	0,0	0,0

Organization: Bàrbara N. Rocha, 2010.

The lowest average temperature (16.7°C) was recorded at the cave entrance, due to the greater environmental fluctuations at this point. The maximum and minimum values for the period were also found there (19.4°C and 14.1°C, respectively), resulting in the greatest temperature range (5.3°C). These variations are the result of incident radiation during the day and heat loss at night, since there is no rock protection at this location.

In the visitable room, located in the middle of the gallery, the average temperature was 17.2°C, with a maximum of 17.5°C and a minimum of 17°C, resulting in a range of 0.5°C. At the end of the corridor, where there are no visitors, the average was 17.1°C, with a maximum of 17.3°C and a minimum of 16.7°C. Despite being furthest from the entrance, this point had higher temperature ranges than the previous point, because it is close to a body of water, which helps to minimize environmental fluctuations.

The modes were close to the means, giving the data stability. This is also borne out by the low variance and standard deviation values.

The relative humidity was even more stable. At the points inside the cave, the air was always saturated, resulting in averages, maxima, minima, modes and medians equal to 100% and, consequently, zero amplitudes, variances and standard deviations.

At the cave entrance, the data showed some variation, although the mean was very close to the saturation point (99.9%) and the mode was 100%. The lowest RH recorded at this location was 68.4%, resulting in an amplitude of 31.6. The variance was 2.8 and the standard deviation was 1.7. This is explained by the location of this collection point in the external environment, subject to variations in the diurnal temperature cycle.

The graph below (figure 5.8.2) shows the variation in air temperature in the external environment (weather station and cave entrance - left axis) and inside the cave (visitable and non-visitable halls - right axis).

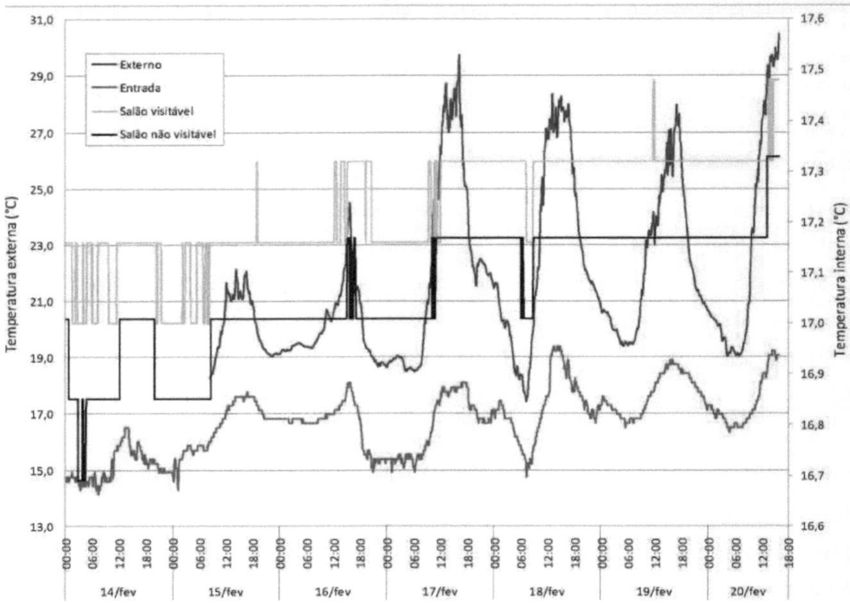

Figura 5.8.2: Air temperature variation (°C) in Santa's Cave.

Organization: Barbara N. Rocha, 2010.

In the cave, the place with the greatest temperature variations was the entrance due to the incidence of direct and diffuse solar radiation at this point. The values clearly follow the cycle between day and night, with smaller thermal amplitudes, but without delays in assimilating the weather outside.

Inside, the variations are small, less than 0.5°C, following the upward or downward trend dictated by the outside environment. However, this assimilation of the weather outside is delayed and cannot be seen every day. The visitors' lounge had temperatures around 0.2°C higher than those at the end of the gallery throughout the period.

On the 14th, the first day of records, a group of three people visited the cave from 1:55 pm to 2:10 pm. At this time, the temperatures inside the cave showed no change, remaining constant at 17.2°C in the visitable room, from 11.30 am until 8.20 pm, and at 17°C in the non-visitable area (figure 5.8.3).

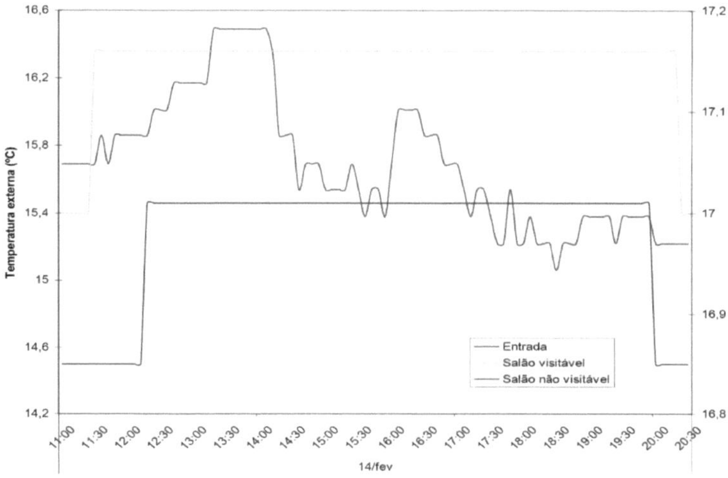

Figura 5.8.3: Hourly variation in air temperature on February 14 in the Santa's Cave.

Organization: Barbara N. Rocha, 2010.

Therefore, even though it has a certain microclimatic stability, small groups of visitors shouldn't cause any changes in the air temperature inside Santa's cave.

The variation in relative humidity can be seen in figure 5.8.4 below.

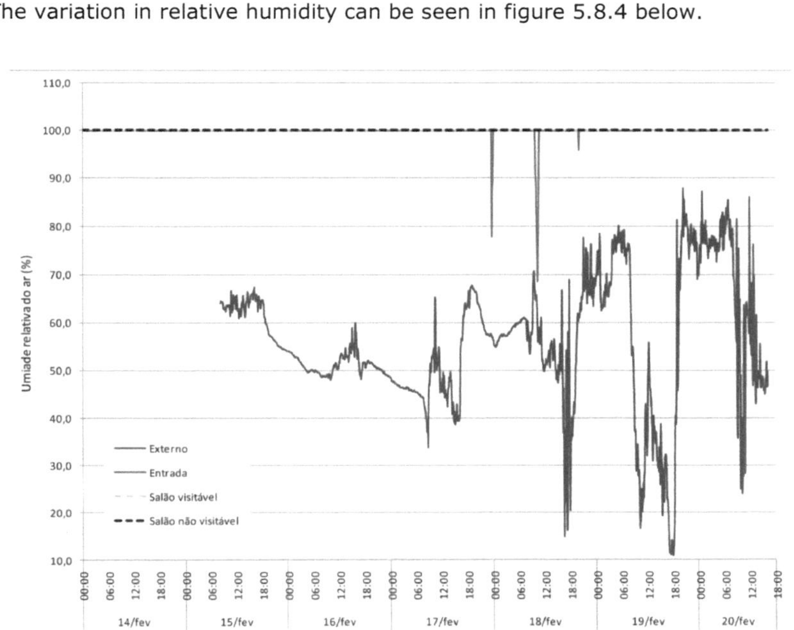

Figure 5.8.4: Variation in relative humidity (%) in Santa's Cave.

Organization: Bárbara N. Rocha, 2010.

Inside the cave, in both the visitable and non-visitable rooms, the relative humidity remained constant at 100%. Variations could only be seen at the entrance to the cave, on February 17th and 18th. The drops in RH at the cave entrance did not follow the variations recorded by the weather station, which showed an increase at those times. The global solar radiation and wind speed data explain these fluctuations (figures 5.8.5 and 5.8.6).

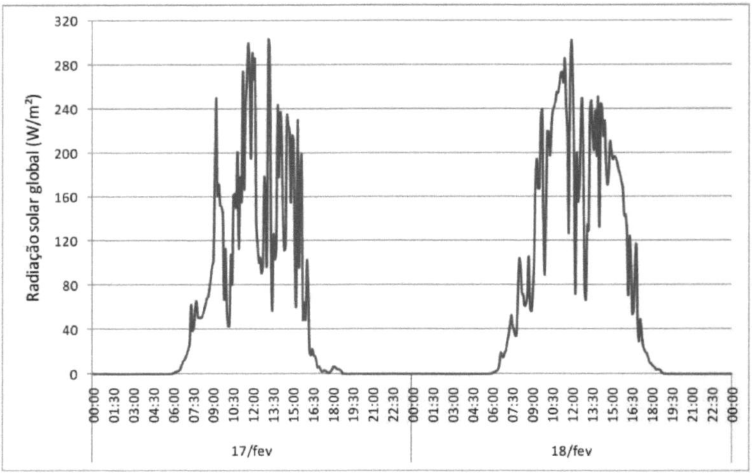

Figure 5.8.5: Variation in global solar radiation at the automatic weather station installed at P.E. Intervales.

Organization: Bárbara N. Rocha, 2010.

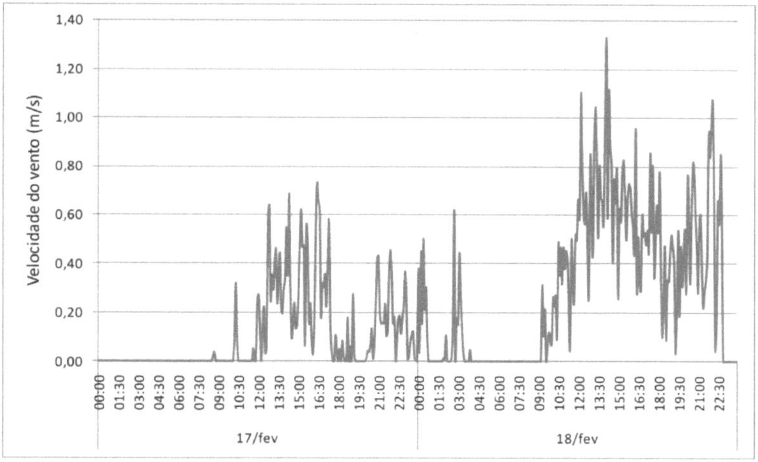

Figure 5.8.6: Wind speed at the automatic weather station installed at P.E. Intervales.

Organization: Bárbara N. Rocha, 2010.

Around 1 p.m. on February 17, 2010, the wind started to pick up speed and was blowing from the southeast. This, combined with the orientation of the cave opening, may have contributed to a renewal of the air at the cave entrance, which may have reduced the relative humidity at this point. As the cave entrance is unsheltered, winds can reach it from any direction.

On the 18th, the humidity dropped around 11am, when the direct sunlight reached its maximum that day. Winds from the south also started to blow more strongly at this time. The variations in RH at the cave entrance can therefore be explained by natural factors.

Figure 5.8.7 shows the carbon dioxide profile of Santa's Cave, taken on February 20 at 9 a.m. in the presence of four people.

In the external environment, the concentration of carbon dioxide is lower (373 ppm) and gradually increases as it enters the duct, reaching 499 ppm at the end of the gallery.

Despite the variation, the difference between the maximum and minimum was small, at 126 ppm, one of the smallest ranges found.

The rate of CO_2 inside the cave is low, as there are practically no speleothems in the cave. The formation of these ornaments is mainly responsible for the release of this gas into the cave atmosphere, as described by Karmann (2003).

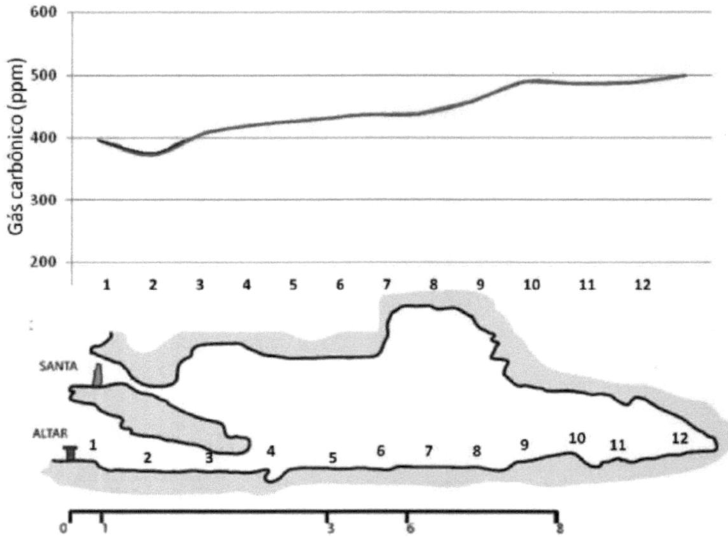

Figure 5.8.7: Carbon dioxide profile of Santa's Cave.

Organization: Barbara N. Rocha, 2010.

In short, no thermal or hygric variations were identified, nor in the concentration of CO_2 , in the Gruta da Santa as a result of tourist visits.

5.9 - Cave of the Minotaur

The Minotaur's cave is made up of several galleries, which gave it its labyrinthine name. It has two main entrances. Near the entrances, the galleries have wide corridors, allowing for a good exchange of energy in the environment. At the end of the cave, the corridors narrow and become more tortuous. A small watercourse runs inside.

Three thermo-hygrometers were installed in this cave, one at the side entrance, another in the middle of the innermost room (visitable room) and the last at the access to the upper gallery, which is closed to visitors. The equipment can be seen in the image below (figure 5.9.1). The absence of a map of the cavity prevented the thermohygrometers from being located. The monitoring period was from February 21 to 28, 2009.

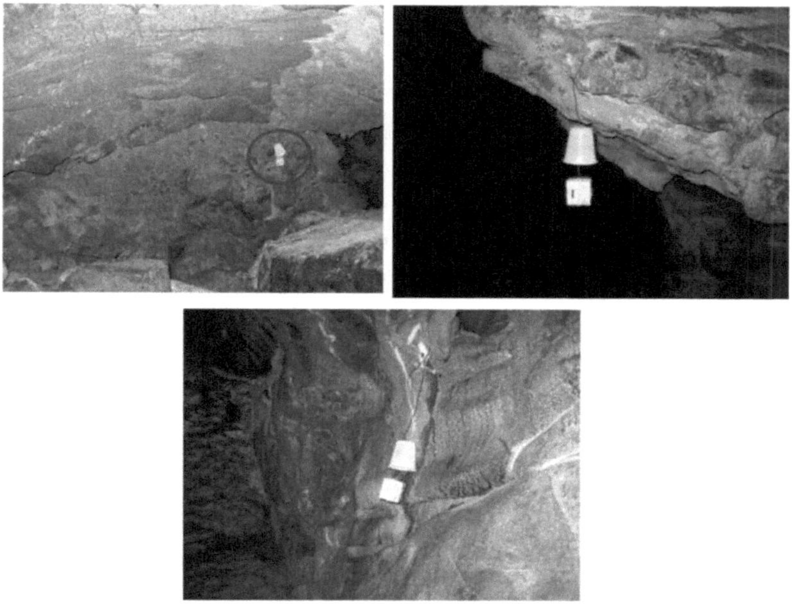

Figure 5.9.1: Photo of the thermohygrometers installed in the Minotaur Cave.

Organization and photographs: Bàrbara N. Rocha, 2010.

The data obtained was processed statistically and can be seen in the table below.

Table 5.9.1: Statistical parameters for temperature and relative humidity humidity in the Minotaur Cave (period from 21 to 28/02/2009).

	Air temperature (^{0} C)			Relative humidity (%)		
	Entry	Salon Visitable	Room not Visitable	Entry	Salon Visitable	Room not Visitable
Average	16,1	17,3	17,5	100,0	90,8	100,0
Maximo	17,8	17,5	17,8	100,0	100,0	100,0
Minimum	14,6	17,2	17,3	100,0	62,9	100,0
Range	3,2	0,3	0,5	0,0	37,1	0,0
Fashion	15,2	17,3	17,5	100,0	100,0	100,0
Variance	0,5	0,0	0,0	0,0	163,6	0,0
Standard deviation	0,7	0,1	0,1	0,0	12,8	0,0

Organization: Bàrbara N. Rocha, 2010.

The entrance to the cave had the lowest average temperature for the period (16.1°C). In the visitable room, located in the innermost part of the cave, the average was 17.3°C and in the non-visitable area (access to the upper gallery) the average was 17.5°C.

Unlike most caves, the maximum was not recorded at the entrance but in the innermost room. Despite being subject to diffuse solar radiation, the side entrance to the Minotaur Cave has well-developed vegetation and a rocky wall on the side that allows the wind to be channeled into the cavity. As a result, the entrance to the cave is quite shaded and allows for the subsidence of cold, denser air on nights with a thermal inversion. The maximum recorded at the entrance was 17.76°C, compared to 17.81°C inside.

The minimum temperature occurred at the inlet, influenced by the physical configuration of this point, and was 14.6°C. The highest amplitude was also at this point, at 3.2°C. Inside the cave, the minimum temperatures were over 17°C and the amplitudes were less than 0.5°C.

As for the modes, at the points located inside the cave this parameter was very close to the average. This information, combined with the low variance and standard deviation values, leads us to conclude that there is thermal stability at these locations. At the entrance, the mode was more variable and the variance and standard deviation were higher.

The relative humidity was 100% throughout the entire period at the entrance to the cave and at the access to the upper gallery. Thus, the averages, maximums, minimums and modes of these points were 100% and the amplitudes, variances and standard deviations were equal to zero.

In the visitors' lounge, the RH varied. The hygrometric average for the period was 90.8%, with a maximum of 100% and a minimum of 62.9%, resulting in a range of 37.1%. This high amplitude resulted in a variance of 163.6 and a standard deviation of 12.8. Despite this, the mode of this point was 100%, indicating that most of the time the air was saturated.

Figure 5.9.2 shows the variation in air temperature inside the cave (right axis) and outside (data from the weather station and the cave entrance - left axis).

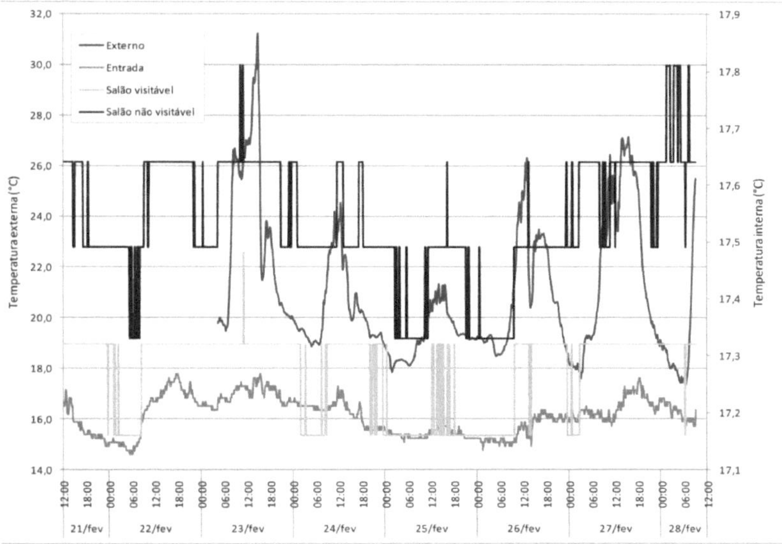

Figura 5.9.2: Air temperature variation (°C) in the Minotaur Cave.

Organization: Barbara N. Rocha, 2010.

At the entrance to the cave, which characterizes the transitional microclimate, the temperature varies throughout the day, with lows at the end of the morning and highs shortly after midday. The temperature follows the variations in the external environment, with reduced thermal amplitudes. However, some variations from the outside cannot be perceived at the cave entrance.

In the non-visitable room, there is a slight variation that slightly follows the tendency of the external environment to change. In the visitable room, at the end of the gallery, the variations are even smaller and the assimilation of the changes in the weather outside can no longer be perceived.

Three groups of tourists visited this cave. One on February 22, consisting of two people, and two on February 23, one with five and the other with four. Figure 5.9.3 shows the hourly variation in air temperature on these two days.

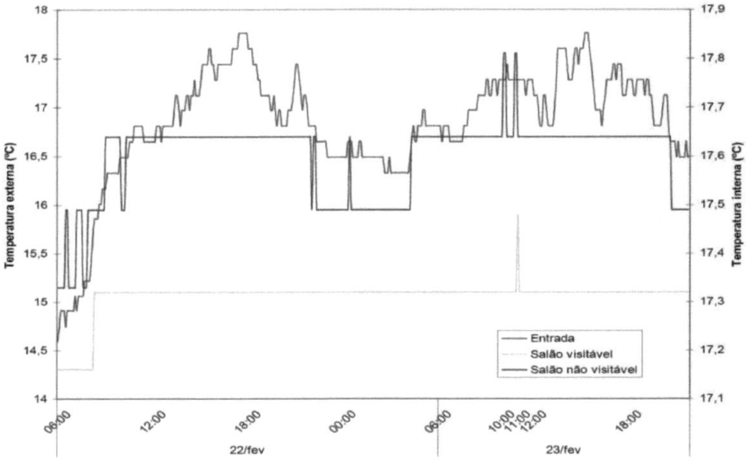

Figura 5.9.3: Air temperature variation on February 22nd and 23rd.

Organization: Barbara N. Rocha.

On the 22nd, no change in air temperature could be seen in the presence of two tourists. On the 23rd, the entry of nine people into the cave from 10:00 am to 11:20 am resulted in changes of 0.2°C in the two rooms analyzed. It took less than ten minutes for the cave to return to a stable state. This change should not have a major impact on the cave, given that it is within the range of the instrumental error and that the greatest changes occurred in a natural situation.

The next graph (figure 5.9.4) illustrates the variation in relative humidity in the Minotaur cave.

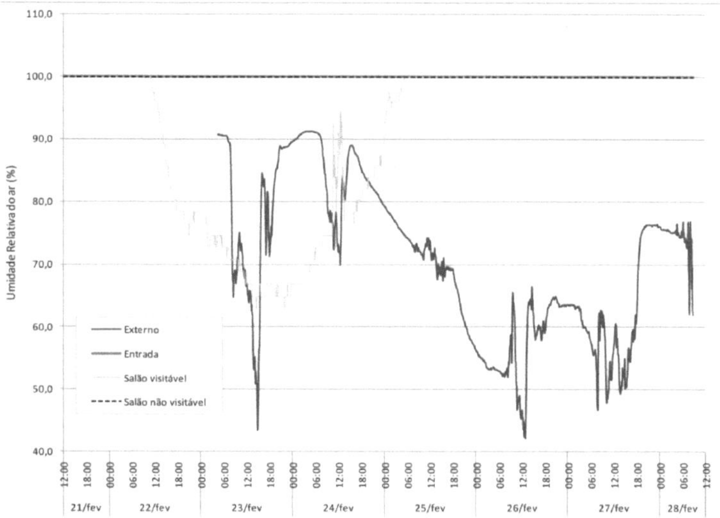

Figura 5.9.4: Variation in relative humidity (%) in the Minotaur Cave.

Organization: Barbara N. Rocha, 2010.

At the entrance to the cave and at the access to the upper gallery, the air was saturated throughout the period, not keeping up with the variations in the external environment. The dense vegetation at the entrance and the presence of a watercourse next to the non-visitable area make these places more humid.

The only variations were in the non-visitable room, located in the final hall. The RH began to fall at 11am on the 22nd. On this day, two people visited the cave at 10.20 am and left at 11.00 am. On the 23rd, nine tourists were in the cave from 10am to 11.20am. At this time, the relative humidity rose and fell again after the visitors left. This variation does not follow the diurnal cycle and may be due to tourist visits. Further studies would be needed to prove this impact.

Figure 5.9.5 shows the carbon dioxide profile of the Minotaur Cave, taken at 9 a.m. on February 28, in the presence of three tourists.

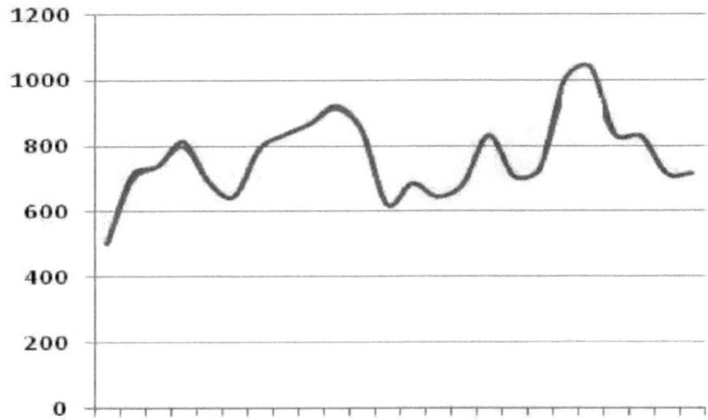

Figure 5.9.5: Carbon dioxide profile of the Minotaur Cave.

Organization: Barbara N. Rocha, 2010.

The concentration of carbon dioxide in the cave varies greatly. In general, wide halls have lower levels of the gas, while narrow passages have higher levels. The lowest value recorded was 502 ppm at the entrance and the highest, 973 ppm, was recorded in the passage to the last room.

Empirically, it was noticed that the narrower halls are poorly ventilated and that CO_2 levels can increase in the presence of tourists. Access to this cave is via a long trail, which makes tourists more exhausted, releasing greater quantities of carbon dioxide into the cave's atmosphere.

From the above, it can be concluded that the presence of tourists in the Minotaur Cave has resulted in variations in the atmosphere of this cave, in the three attributes analyzed (temperature, relative humidity and carbon dioxide rate). The labyrinthine shape of the cave hinders gas exchange between its halls and the external environment. In future studies, it is recommended that this cave be studied for a longer period of time in order to better understand the magnitude of these impacts.

5.10 - Summary of results

The aim of the microclimate monitoring was to check whether or not atmospheric parameters were amplified by the flow of visitors and the climatic conditions around the group of caves.

Table 5.10.1 summarizes the physical aspects of caves that influence their microclimate.

Table 5.10.1: Physical aspects of cavities that influence their microlime.

	Physical aspects									
	Halls		Presence of a river		Presence of skylights		Extension		Mouth	
Cave	Wide	Narrows	Yes	No	yes	No	Small	Medium/large	Wide	Narrow
Detrâs		x		x	x		x			x
Fire	x		x			x		x	x	
armadillo	x		x		x		x		x	
Boys		x		x	x		x			x
Jane Mansfield	x		x			x		x		x
Cleft	x		x		x			x	x	
Maozinha	X			x		x	x		x	
Santa	X		x			x	x		x	
Minotaur	x		x			x		x	x	

Organization: Barbara N. Rocha, 2010.

Halls or corridors where at least two people of average height and weight can walk side by side were considered wide. In the Meninos and Cipó caves, all the corridors are narrow.

Most of the caves studied have watercourses inside. When the rivers are flowing, as is the case in Gruta do Fendao and Jane Mansfield, and the cave has skylights, air circulation is facilitated, a fact that could be empirically proven.

In caves with narrow mouths, where it is necessary to crouch down to enter them, the influence of external climatic variations tends to be less; while in small caves (less than 100m in length) a greater variation in climatic attributes is expected.

The microclimate of Detràs Cave follows the variations in the external environment to a lesser extent. This is due to its small size and the presence of skylights. However, this cave showed low-magnitude impacts (variations lower than those that normally occur over the course of a day) on its temperature as a result of visits by a small group of tourists.

In the Cave of Fire, increases in air temperature and carbon dioxide concentrations were detected in the presence of tourists in the two halls studied. This cave has no skylights, which, together with its medium length, prevent the atmosphere inside from "seeing" the variations in the external environment.

Toca dos Meninos is a small cave. For this reason, up to the middle of the gallery, the climate in this cave follows the variations in the external environment. In the final room, which is more confined, the temperature tends to remain stable.

The atmosphere of the Tatu cave follows the variations in the external environment. Its shape, with a wide entrance and corridors, a small area and the presence of a river and skylights, increases the capacity for gas exchange with the outside environment.

The Fendao Cave also has a high capacity for energy exchange with the external environment, due to its large halls with skylights and a flowing river.

The variations in temperature and relative humidity in Jane Mansfield Cave are due to the high energy exchange provided by the flowing river. This cave did not show any changes in its atmosphere during periods of tourist visitation.

Under normal conditions of visitation (without the use of carburetors), no changes were measured in the atmospheric parameters of the Maozinha Cave during periods of visitation. Its small size and wide mouth allow for greater energy exchange with the external environment.

The Santa's cave shows stability in the microclimatic attributes measured. Only in the areas close to the cave entrance can external climatic variations be perceived.

The Minotaur Cave has narrow corridors and a labyrinthine shape, making it difficult to exchange energy with the external environment. For this reason, it showed changes in all the climatic parameters analyzed during tourist visits. The temperature rose in both halls, the relative humidity fell in the visitable hall and the CO levels$_2$ were variable in almost every profile.

6. Conclusions and final considerations

This work has proved to be of fundamental importance in understanding the microclimate of cave environments. In general, caves have a high relative humidity, reaching saturation, and a temperature that tends to remain stable as you enter the caves. The concentration of carbon dioxide is higher inside caves than outside.

By correlating the data obtained with the shape of the caves, it can be concluded that caves with large entrances and skylights have a microclimate that is still governed by the dynamics of the external environment. Caves with smaller openings do not "see" the outside atmosphere.

The presence of flowing rivers and skylights proved to be more decisive factors in the variation of the microclimate inside caves than the length of the cave, the breadth of the halls and the width of the mouth.

All the caves with a flowing river had a variable (non-stable) microclimate. It can therefore be concluded that water is an important circulation mechanism in cave environments.

The caves that have skylights also have a variable microclimate, as there is gas exchange with the external environment in these caves.

Variations in the microclimate of the caves studied as a result of tourist visits generally occurred in dry caves, with narrow corridors and entrances and no skylights. These characteristics hinder air circulation and the consequent energy exchange between the external and internal environments. The caves that showed impacts were the Minotaur, Detràs, Fogo, dos Meninos and Maozinha (the latter due to the use of carburetors).

In small caves, it was believed that gas exchange with the external environment would be facilitated, preventing the stabilization of the microclimatic parameters measured, especially air temperature. However, the smaller caves were the ones that showed impacts from tourism, showing that the size of the cave is not an important factor for air circulation inside caves.

Air circulation is facilitated in caves with flowing rivers, large halls, wide entrances and corridors and the presence of skylights. This generates greater variations in the attributes of the microclimate, making it difficult to generate impacts on the environment as a result of tourism.

Tests carried out in the Maozinha cave have also shown that the use of carburetors significantly alters the concentrations of carbon dioxide in the cave atmosphere. A single device switched on was enough to increase CO_2 levels in the cave by 2.5 times.

It is worth remembering that the methodology for this study does not allow us to understand the natural seasonal changes in the climate of the cave environment. Therefore, in some cases, it is not possible to conclusively state whether the changes found are the result of anthropogenic presence or natural factors.

It should also be noted that the visitor tests were carried out with small groups of visitors. Therefore, conclusions about the impacts of medium and large groups, which bring a greater energy input to the interior of the cavities, cannot be extended.

The results of this work allow us to accept, in part, the hypotheses formulated.

1) The microclimate of caves, in a natural situation, shows delays in the assimilation of external atmospheric conditions, as well as a reduced temperature range. However, in some caves, especially those with flowing rivers, the air temperature can vary in the aphotic zones, but with smaller amplitudes than in the external environment.

2) Tourist visits, through the use of artificial lighting and human body heat, can alter the air temperature in some caves. However, changes in temperature do not result in variations in relative humidity.

3) Cave tourism using carburetors does increase the concentration of carbon dioxide in the atmosphere.

Finally, we believe that this work can contribute to a better understanding of this unique and understudied environment which is caves. It should be noted, however, that this research does not exhaust the subject. Much remains to be studied if we are to understand the rhythm of the cave microclimate. The time interval used to record the climate attributes in this study (one week) does not allow us to understand the seasonality of the climate in the caves. Studies with analysis over a longer period of time are suggested for future research. Other studies could also seek to identify the support capacity of caves, considering their microclimatic aspects.

7. Bibliography

ALEY, T. De volta à primeira casa: o mundo subterrâneo das cavernas In: **Ciência e Futuro 1997**. Sao Paulo: Ed. Melhoramentos, 1997.

ARMANI, G., TAVARES, R. e ROCHA, B. N. Caracterizaçao pluvial da Baixada Santista como subsidio à prevençao de desastres naturais. Proceedings of **the XI Brazilian Symposium of Applied Physical Geography**. Sao Paulo, 2005.

ARMANI, G., TAVARES, R. and ROCHA, B. N. The influence of sea and land breezes on the genesis of rainfall in the Cubatao-Sao Paulo (SP) transect. Proceedings of **the XIV Brazilian Congress of Meteorology**. Florianópolis, 2006.

BRAZIL. **Decree No. 750.** 1993.

BRAZIL. **National Environmental Policy**. Law No. 6983 (1981).

BRAZIL. **CONAMA Resolution No. 001.** 1986.

BRAZIL. **CONAMA Resolution No. 347**. 2004

CARVALHO, S. M. Study of underground microclimate: the example of the Olhos D'àgua Cave - Castro (PR). In: ZAVATTINI, J. A. **Estudos do clima no Brasil**. Sao Paulo: Alinea editora, 2004. p. 212-213.

CONTI, J. B. **Secondary circulation and orographic effect in the genesis of rainfall in the northeastern region of São Paulo**. Sao Paulo, 1v. Thesis (Doctorate) - Faculty of Philosophy, Letters and Human Sciences, University of Sao Paulo. Sao Paulo, 1975. 82p.

GEIGER, R. **Manual de microclimatologia: o clima da camada de ar junto do solo**. 4. ed. Lisbon: Calouste Gulbenkian Foundation, 1961. 556p. GILLIESON, D. **Caves: processes, development, management**. Blackwell, 1996. 324p.

GUTJAHR, M. R. **Criteria related to the climatic compartmentalization of river basins: the Ribeira de Iguape River Basin**. Sao Paulo, 1v. Dissertation (Master's Degree) - Faculty of Philosophy, Letters and Human Sciences, University of Sao Paulo. Sao Paulo, 1993.

KARMANN, Ivo. Water cycle: underground water and its geological action. In: TEIXEIRA, W., **Deciphering the Earth**. Oficina de textos: Sao Paulo, 2003.

LINAN, C., VADILLO, I e CARRASCO, F. La concentración de CO2 del aire em la Cueva de Nerja (Màlaga, Espana). **Hidrogeologia y Aguas Subterrâneas, 18**, 2006. P. 357-364.

LINO, C. F. **Cavernas: o fascinante Brasil subterrâneo**. Sao Paulo: Gaia, 2001.

LINO, C. F. and ALLIEVI, J. **Brazilian Caves**. Sao Paulo: Ediçoes Melhoramentos, 1980. 168p.

LONGHITANO, G. A., ROCHA, B. N. and ÂNGELO FURLAN, S. Microclimatic characterization of the Gruta Colorida - Intervales State Park, SP. In: **Proceedings of the VII Brazilian Symposium on Geographical Climatology**. Rondonópolis, 2006.

MILANESI, M. A. The island of Sao Sebastiao in the state of Sao Paulo: rainfall studies. Proceedings of **the VI Brazilian Symposium on Geographical Climatology**. Aracaju, 2004.

MONTEIRO, C. A. de F. **Climate dynamics and rainfall in the State of São Paulo**. Atlas. Sao Paulo, Institute of Geography/USP, 1973. 129p.

MONTEIRO, C. A. de F. **Teoria e clima urbano**. Sao Paulo, 1v. Thesis (Livre docência) - Faculty of Philosophy, Letters and Human Sciences, University of Sao Paulo. Sao Paulo, 1975.

PALOMBA, V. **Proposta per la fruzione naturalistica di un ambiente ipogeo neo parco fluviale del revere**. Università Degli Studi di Perugia - Facolta' di Scienze, 1995.

PELEGATI, C. H. G. The orographic effect on the Mogi-Bertioga profile. **Proceedings of the XI Brazilian Symposium of Applied Physical Geography**. Sao Paulo, 2005.

ROCHA, B. N., LONGHITANO, G. A. and ÂNGELO FURLAN, S. Climatic-faunalistic survey of the Colorful Cave in Intervales State Park. In: **Proceedings of the IV Brazilian Congress of Biometeorology**. Ribeirao Preto, 2006.

RODRIGUES, C. The geosystemic theory and its contribution to geographic and environmental studies. **Revista do Departamento de Geografia**, Sao Paulo, n. 14, p. 69-77, 2001.

ROSS, J. L. S. The morphogenesis of the Ribeira do Iguape river basin and environmental systems. **Revista GEOUSP - Espaço e Tempo**, Sao Paulo, n. 12, p. 21-26, 2002.

ROSS, J. L. S. and MOROZ, I. C. **Geomorphological map of the State of São Paulo**. Scale 1:500000. Sao Paulo: Technological Research Institute, v. I/II, 1997.

SANCHEZ, L. E. **Environmental impact assessment: concepts and methods**. Sao Paulo: Oficina de Textos, 2008. 495p.

SANCHEZ-MARTOS, F., CALAFORRA, J. M., FERNANDEZ-CORTES, A. and GONZALES-RiOS, M. J. Experiencia de visitas masivas a cavidades en condiciones naturales: la Cueva del Agua de Iznalloz (Granada). **Geogaceta, 31**, 2002. P. 23-26.

SANO, N. N. **Comparative study of visitation management in the Alto Ribeira (PETAR) and Intervales (PEI) Tourist State Parks**. Master's degree dissertation. Sao Paulo: USP, 2007. 100p. (+ annexes).

SANT'ANNA NETO, J. L. **As chuvas no Estado de Sâo Paulo: contribuçâo ao estudo da variabilidade e tendência da pluviosidade na perspectiva da análisis geogràfica**. Sao Paulo, 1v. Thesis (Doctorate) - Faculty of Philosophy, Letters and Human Sciences, University of Sao Paulo. Sao Paulo, 1995. 201p.

SANT'ANNA NETO, J. L. **Climatic rhythm and the genesis of rainfall in the São Paulo coastal zone**. Sao Paulo, 1v. Dissertation (Master's Degree) - Faculty of Philosophy, Letters and Human Sciences, University of Sao Paulo. Sao Paulo, 1990.

SAO PAULO (State). **Intervales State Park: environmental management plan - phase 1**. 1994.

SAO PAULO (State). **Intervales State Park: management plan**. 2009.

SCALEANTE, J. A. B. **Evaluation of the impact of tourist activities in caves**. Master's thesis. Campinas: UNICAMP, 2003. 70p. (+ annexes).

SOTCHAWA, V. B. The study of geosystems. In: **Revista IG-USP** (Cadernos Métodos em Questao), n.16. Sao Paulo, 1977.

VERiSSIMO, C. U., SOUSA, A. E. B. A. de, RICARDO, J. M., BARCELOS, A. C., NOGUEIRA NETO, J. de A., REIS, M. G. M. Microclimate and speleotourism in the cave of Ubajara, CE. In: **Anais... 27th Brazilian Congress of Speleology**. Januâria, 2003.

VIANA JUNIOR, O. **Hydrochemistry, hydrology and isotopic geochemistry (O and H) of the autogenic vadose percolation facies, Santana Cave, municipality of Iporanga, State of São Paulo**. Master's thesis. Sao Paulo: Institute of Geosciences/ USP, 2002. 113p.

6.1 - Sites consulted

USP Chemistry Department. Available at: http://www.cdcc.usp.br/quimica/ciencia/cavernas.html. Accessed on October 16, 2009.

Department of Geography, FFLCH/ USP. Available at: http://www.geografia.fflch.usp.br/mapas/Atlas_Intervales/. Accessed on October 15, 2009.

ECOTUR Magazine. Available at:

http://www.revistaecotour.com.br/novo/home/default.asp?tipo=noticia&id= 1247. Accessed 28/10/2009.

8. Annexes

8.1 - Cave visit questionnaire

Cave: ________________________________

Date	Entry times	Departure times	Number of visitors	Lighting sources

MIX
Papier aus verantwortungsvollen Quellen
Paper from responsible sources
FSC® C105338

Printed by Books on Demand GmbH, Norderstedt / Germany